*INTERNATIONAL SERIES OF MONOGRAPHS ON
PURE AND APPLIED BIOLOGY*

Division: **PLANT PHYSIOLOGY**

GENERAL EDITORS: P. F. WAREING and A. W. GALSTON

VOLUME 4

THE
FLOWERING PROCESS

A

OTHER TITLES IN THE PLANT PHYSIOLOGY DIVISION

TITLES IN THE BOTANY DIVISION

OTHER DIVISIONS IN THE
SERIES ON PURE AND APPLIED BIOLOGY

BIOCHEMISTRY

MODERN TRENDS IN PHYSIOLOGICAL SCIENCES

ZOOLOGY

This birch tree in Tübingen, Germany, illustrates well the response to day-length which is the principle topic of this book. The street lamp at the lower right has a reflector on top, so that it casts its light only downward and outward. The leaves illuminated by the light at night have responded to this artificially extended day by not going into dormancy, even though it is the first week of November, and Tübingen has already had some severe frosts. Most of the leaves that were in the dark have fallen from the tree some weeks ago. The number and green color of the remaining leaves are closely proportional to the light intensity they receive from the lamp, illustrating a quantitative aspect to this response to a time factor (the length of day).

THE
FLOWERING PROCESS

BY

FRANK B. SALISBURY

Professor of Plant Physiology
Colorado State University

PERGAMON PRESS

OXFORD · LONDON · NEW YORK · PARIS

1963

PERGAMON PRESS LTD.
Headington Hill Hall, Oxford
4 and 5 Fitzroy Square, London, W.1

PERGAMON PRESS INC.
122 East 55th Street, New York 22, N.Y.

GAUTHIER-VILLARS ED.
55 Quai des Grands-Augustins, Paris, 6e

PERGAMON PRESS G.m.b.H.
Kaiserstrasse 75, Frankfurt am Main

Distributed in the Western Hemisphere by
THE MACMILLAN COMPANY · NEW YORK
pursuant to a special arrangement with
Pergamon Press Inc.

Library of Congress Catalog Card Number 63-19739

Set in Monotype Times New Roman 10 on 12 pt.
and printed in Great Britain by
THE BAY TREE PRESS, STEVENAGE, HERTS.

TO
JAMES BONNER
FOR INVOLVING ME IN
THIS FASCINATING SUBJECT

CONTENTS

PREFACE

THE principal purpose of this book is to discuss in its biological framework, the conversion from the vegetative to the reproductive state in higher plants. There are two aspects to the study of this conversion; first, the changes within the plant which lead to the conversion, and second, the conversion itself. The first of these aspects has been studied most and is emphasized.

The extent of scientific development in this field is quite amazing. Probably only a small portion of the world's population is aware of this rather isolated branch of science, but a complete collection of papers relating to the physiology of flowering would fill a rather impressive bookshelf. It would be fairly easy to find 1000 such papers. Thus a straightforward complete review of this work would probably result in a very thick volume. Luckily, space allotments from the Publisher saved the author's time and patience from being put to such a test. A complete summary volume would be of unquestioned reference value to science, but unless the author were gifted, such a condensed recounting of experimental work would quite probably make for very dry reading. All of this poses a dilemma for an author: he can try to cover the field and probably lose his reader in the mass of conflicting and often unrelated facts, or he can concentrate on certain aspects of the physiology of flowering and thereby slight other aspects which may be equally interesting and important. I decided on the second approach.

The book is addressed to graduate students and others who might be interested in the topic presented approximately at the graduate level. It is assumed that the reader has a good background in some phase of biology but that his acquaintance with the physiology of flowering is rather cursory. It was my intention to discuss broad aspects of the topic in the first four chapters, but in some respects these became as specific as the ones which follow. I feel that they will provide a good introduction to the last part of the book for the student who already has some knowledge about the flowering process, but the real beginner might want to return to Chapters 2, 3, and 4 after the other chapters have been completed.

In the last six chapters I gave in without reservation to the temptation to discuss in some detail my own main interest, relying heavily upon personal research experience. This interest is in the sequence of biochemical and biophysical events which take place within the plant, beginning first with response to the environmental stimulus imparted by the relative length of day and night and culminating in the production of flowers by the plant primary shoot meristems. Although many species are mentioned, the theme of the narrative always centers around the cocklebur.

This is not because this plant is highly "typical". The converse is probably true, and the principal atypical response of this plant, flowering after exposure to a single long dark period, makes it an ideal research plant. Thus my preoccupation with this species as a "type", even though it is not exactly "typical", is based upon its nature as an experimental object. This nature readily allows the experimenter to think of the flowering process as a series of catenary events, each bearing some time relationship to the single "inductive" dark period. Other plants are now known to be equally well suited, but our experience with them is not yet so extensive as that with the cocklebur.

In an early version of the manuscript, the book was addressed largely to high school teachers of biology. There was one aspect of this early approach which appealed to me very much: the flowering process is a fairly good summary of biology in general. This is discussed briefly in Chapter 1, and it is hoped that the idea is evident throughout the book. The breadth of such an isolated topic is quite impressive, and this breadth must surely be typical of what one might find upon intensive study in virtually any "narrow" field. There is a unity in science, and the specialist who would really specialize will find more and more that he must be a general practitioner.

Since it seemed desirable to avoid the style of a literature review, an effort was made to reduce the number of references in the text to a bare minimum. This is possible only because a number of excellent reviews have been written in recent years. These are listed in the bibliography, and section headings often refer the reader to a number of them. Such references in section headings were chosen according to my impressions about the reviews with which I am most familiar.

Many of the books and reviews are at least as broad as the present volume and could be used as references in virtually all sections. The interested reader who wants to see original papers can find references according to topic in nearly any of these recent books or reviews (see especially 3, 9, 11, 14, 20, 21, 22, 25, 26, 28, or 32). Actually, nearly all work not directly cited in this book is documented in my own review (32). The printed report of the recent symposium in Australia will contain very recent references, and I am indebted to Dr. Erwin Bünning for showing me some of these manuscripts. My last revision was strongly influenced by them.

In spite of this approach to literature citation through reviews, it was felt that more direct reference should be made in cases where work (rather than well-known conclusions) is specifically mentioned, using the name and location of the investigator. In these cases a recent pertinent paper is cited. Figures copied from published papers are also acknowledged in the figure captions, giving further specific references. Table 7–1 contains references to a number of papers which are either quite recent or not easily found in the reviews. There is also a considerable amount of unpublished work which is discussed in the book. Usually this is apparent from the figure headings.

The manuscript was used in an early duplicated draft in an advanced plant physiology class at Colorado State University during the spring of 1962. As a result of discussions in this class, many of the ideas now incorporated into the text developed, and a number of experiments were performed. Thus I am indebted for both intellectual and material help to the members of this class: Charles Curtis, Lee Eddleman, Nagah Karamani El Sayed, James Gary Holway, Deogratias Lwehabura, Oscur Schmunk, and James Whitmore.

After arriving in Tübingen (in August, 1962, for a sabbatical year), the manuscript was almost completely rewritten. Drs. Arthur Galston, Anton Lang, Jan Zeevaart, and Phillip Wareing had read the duplicated version, and their comments contributed much to the rewriting. Drs. Erwin Bünning and Lars Lörcher also read parts of the manuscript and made valuable suggestions. During the rewriting, Drs. Galston and Wareing supplied immeasurable help by reading and commenting on the Tübingen version. I am also deeply indebted to my assistant, Jean Livingston, and my graduate student Carol Pollard, who answered my many mailed requests to Colorado, sometimes by performing experiments to answer questions that kept coming up. My wife, Marilyn, was indispensable during this period, since she typed the rewritten manuscript to send to Drs. Galston and Wareing.

The following secretary-technicians have helped with clerical and experimental aspects of our cocklebur research in Colorado since 1955: Pauline Christiansen, Anita Brooks, Joan Maxwell, Annette Hullinger, Marjorie Smith, Katherine Kline, Marilyn Young, and Sandra Howard (who typed one complete version of the manuscript). My colleagues in the Department of Botany and Plant Pathology, especially Dr. Cleon Ross who works on biochemical aspects of cockleburology, have been helpful in many ways. Dr. Ross read and commented on the final manuscript. My graduate students have contributed materially to the original work, some of them (Walter Collins, Leona Harrison, and Carol Pollard) by dissertation work on the physiology of flowering, and all of them (Edward Olsson, Robert Mellor, Merrill Ross, and George Spomer) by ungrudgingly helping with all-night experiments.

I would especially like to express appreciation to Colorado State University, the National Science Foundation, and the National Institutes of Health for providing facilities and financial support for our cocklebur research and for preparation of this manuscript.

FRANK B. SALISBURY.

Botanisches Institut
Tübingen, West Germany

Note — Although decimal points throughout this book are given according to American practice (and also the spelling, since the author is a U.S. citizen), on several of the diagrams raised points appear in the decimals. It is hoped that this small inconsistency will not mislead the reader.

CHAPTER 1

FLOWERING IN ITS BIOLOGICAL FRAMEWORK

SOME of the most general, and indeed the most exciting aspects of biology are an integral part of the flowering process, and most of the basic fields are represented to a greater or lesser degree. Any study of plant or animal function is physiology, and so the discussions to follow will emphasize physiology. Of course any function is dependent upon some entity or structure, and in the study of flowering we are concerned with the origin of structure itself. Thus we approach the fields of anatomy or morphology. Many plants flower in response to some change in the environment, so the topic bears a valid relationship to ecology. Different kinds of plants respond in different ways, and as one tries to organize these responses according to type, one does work not too unrelated to that of the taxonomist. The flowering response is inherited, and it is possible to study its genetics; indeed, flowering involves the response of the genes and their products to the environment, and study of such things lies in the new field of molecular biology. If there were space, one could also discuss certain applied aspects of flowering in the fields of horticulture and agronomy. Obviously, if flowering could be controlled, agriculture could be revolutionized.

It is amazing how a study in depth of any topic in biology may cut across nearly the entire field of biology itself. The process of flowering is certainly no exception, although there are aspects which are not encountered, such as nerve or muscle function. Certain rather unlikely subjects such as paleobotany or evolutionary mechanisms do bear a relationship to flowering, although we will not have much to say about them here.

We will not approach the flowering process by studying its relationship to each of the traditional fields listed above. Rather, we will keep in mind five general biological areas:

1. Diversity and Uniformity of Biological Material

1

2. Response of an Organism to its Environment
3. Biological Timing
4. Biochemistry
5. Morphogenesis or the Origin of Form

A brief discussion of these five topics now will serve as an introduction to the more detailed discussion of the flowering process which follows. We think of flowering in terms of component steps or events which are taking place within the plant and which ultimately lead to the formation and development of flowers. The whole point of the first topic is that these steps may vary considerably from one species to another. Thus in discussing the last four topics (and in the last six chapters of this book) we shall consider the steps primarily as they are thought to occur in our "type" plant, the cocklebur, although deviations will often be mentioned.

1. *Diversity and Uniformity of Biological Material*

In considering this topic one cannot help feeling somewhat like a pendulum. It is quite obvious that the world of living things consists of a myriad of diverse forms. The list of known species extends into the millions and the diversity is enormous. Consider the protozoa, jelly fish, sponges, flat, round, and segmented worms, starfish, shell fish, snails, shrimps, finned fish, lizards, birds, mammals, and all the other sundry groups of animals. Then think of the bacteria, many kinds of algae, fungi, mosses, liverworts, ferns, conifers, and flowering plants. The taxonomist estimates that we shall one day know three to five million kinds of insects alone. Thus on one swing of the pendulum we are fully aware that there are many kinds of living things.

Yet the significant generalization of modern biology is that all of these various organisms have a number of important and basic functions in common. This is most striking when one considers the biochemistry of the cell. Respiration, for example, proceeds along essentially the same metabolic pathways in all living things, and this also seems to be true for many other processes such as protein synthesis, fat metabolism, etc. So the other end of the pendulum's swing is the concept that living things are really all very much alike. Is this true in the flowering process? At this stage of the game we simply do not know. Some workers have assumed that it was true —

that the flowering process is essentially the same in all flowering plants with only slight modifications which apparently lead to a diversity of response. In my opinion it is too soon to draw this conclusion. It has in the past led to application of findings obtained with one plant to understanding of flowering in another — and subsequent work has frequently failed to support this.

We shall see in the next chapter that the diversity of response in flowering is very great. If we want to make the classification scheme complex enough, we can probably produce a separate category for each species or variety. In many cases these differences are quite striking. A short-day plant is inhibited in its flowering by a brief light interruption of the dark period. A long-day plant is promoted in its flowering by the identical treatment. In one short-day plant far-red light is without effect (or promotes) during the dark period; in another it inhibits flowering. The pendulum should be allowed to swing far to the diversity side, and Chapter 2 is written to try to push it far in that direction.

But it must also be allowed to swing back to the uniformity side. If there is any sort of natural relationship among the flowering plants, as modern biochemistry implies, why shouldn't there be some basic, common underlying mechanisms in the flowering process? There are at least two excellent reasons to think that this is the case. The pigment system which switches the plant's metabolism from the light to the dark status seems to be common to all higher plants — certainly to the ones which we will be discussing. Furthermore, there is evidence from grafting experiments that the flowering hormone itself is the same in species and varieties which in other respects show opposite responses.

Is the apparent diversity of response really only a matter of slight modification of a common basic mechanism? Or have the modifications become so extensive that we should not think in terms of a single mechanism but rather of a number of fundamentally different mechanisms which do happen to be similar in certain respects? Much more research is required before these questions can be answered, and so for the present we can only let the pendulum swing freely while we wait for the facts to come in. The situation is, at any rate, common to most of biology. We are impressed by the uniformity, but the diversity is becoming more and more interesting.

FIGURE 1–1

In order to show the basic response of a number of species to day-length, seeds were planted in the spring of 1962 by Mohamed N. K. El Sayed of the advanced plant physiology class at Colorado State University, and half of the plants of each species were placed under a light-proof box every day at approximately 4.00 p.m. and removed the following morning around 8.00 a.m., while the remaining plants were left under the long-day conditions of our cocklebur greenhouse (about 20 hr of light— see Chapter 5). At various times after planting, as indicated by figures in each picture, the plants were photographed. Scientific names are given in the appendix. Figure A is an example of a day-neutral plant; Figs B to F are absolute short-day plants; Fig. G is nearly an absolute short-day plant, although flowers can also be seen under long-day conditions occasionally; Fig. H is at best only a quantitative short-day plant (see Chapter 2), since it flowers on both long and short days, but faster on short days; and Figs. I to L are absolute long-day plants. Note in most photographs the strong effects of day-length upon vegetative growth as well as flowering. In many cases, exposure of the plants to the day-length which causes flowering from the time they first emerge as seedlings produces flowers on such small plants that the resulting examples are not very typical of flowering plants in nature (e.g. Figs. B, C, D, F, and K). Thus in Figs. E and I, plants were held under non-inductive conditions for a few weeks before they were induced to flower.

FIG. 1–1, A

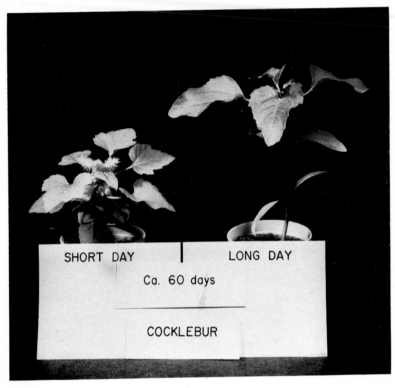

FIG. 1-1, B

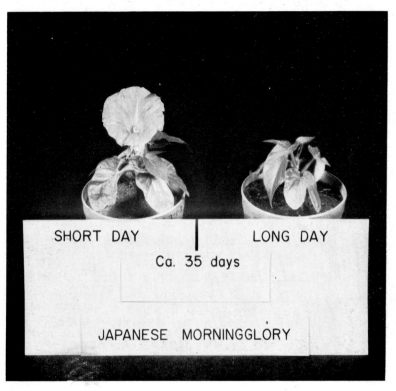

Fig. 1–1, C

Fig. 1–1, D

Fig. 1–1, E

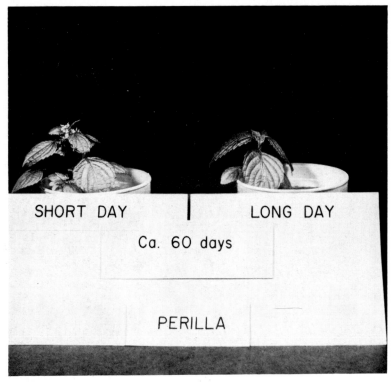

SHORT DAY | LONG DAY

Ca. 60 days

PERILLA

Fig. 1–1, F

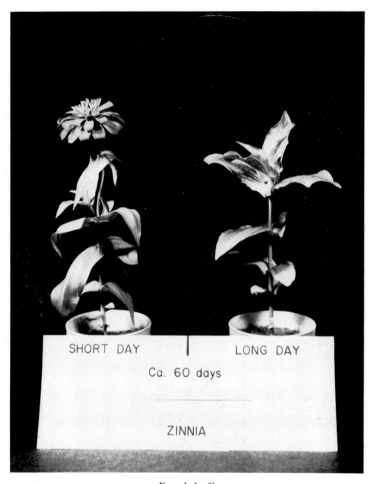

SHORT DAY | LONG DAY

Ca. 60 days

ZINNIA

Fig. 1–1, G

Fig. 1–1, H

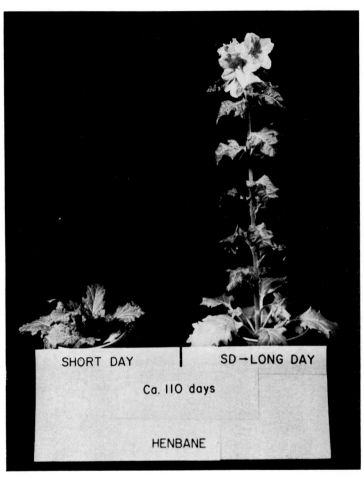

Fig. 1–1, I

FIG. 1-1. J

Fig. 1–1, K

SHORT DAY | LONG DAY
Ca. 35 days
RADISH

Fig. 1–1, L

2. *The Response of an Organism to its Environment*

In our "type" plant, the cocklebur, the fundamental response to environment is a response of the leaf to an uninterrupted dark period which exceeds about 8 hr 20 min. Given such a dark period, the plant flowers; on shorter dark periods the plant remains vegetative. Of course there are other effects of environment: temperature must be right, adequate soil nutrients and water must be available, and if the dark period is to be highly effective, it must be preceded and followed by exposure of the plant to high intensity light. Obviously the response to environment is a matter of physiology, but it can nevertheless be considered in an ecological sense (see Chapter 3).

The whole modern study of the flowering process was initiated quite recently (1920) by the discovery that flowering in many plants is an environmentally conditioned response. W. W. Garner and H. A. Allard, working at the United States Department of Agriculture Plant Industry Station at Beltsville, Maryland, wondered about the peculiar flowering habits of two economically important species. A variety of tobacco, called Maryland Mammoth, grew 10 ft tall during the summer months at Beltsville, but failed to flower and set seed. Transplanted as cuttings or root-stocks into the greenhouse, plants would flower in winter when they were less than 5 ft tall. A certain variety of soybean, when planted successively at various times throughout the spring, tended to come into flower on the same summer date regardless of the planting time. This variety (and some others as well) would flower in winter in the greenhouse even when the plants were very small. Obviously there was something about winter greenhouse conditions which seemed to cause flowering in these two species.

Garner and Allard first tested effects of light intensity, temperature, and available soil moisture and found no definite effect on flowering. Then, almost reluctantly, they tested the effects of day-length, by extending it with artificial light or shortening it by placing plants in cabinets. They were thus able to show that flowering occurred in these two plants when days were shortened — regardless of other environmental conditions (providing it was not too hot, dry or shady for survival). They called the class of plants that responds in this way short-day plants. With other species, long days resulted in flowering (long-day plants), while flowering occurred in some plants on any day-length (day-neutral plants). This basic response is

c

illustrated, with some plants commonly used in such studies, in the photographs of Fig. 1–1. Garner and Allard called their newly discovered phenomenon *photoperiodism*. In later experiments of these and other workers, it was found that many plants respond more to the night-length than to the day-length; thus the term *photoperiodism* is not entirely accurate, but usage has made it secure.

It was known by nineteenth century farmers in the United States that winter wheat, which usually flowers in the spring after being planted the previous fall, will flower even if it is planted in the spring, providing that moist seeds have been exposed to low temperatures for a few weeks. This flowering response was also studied intensively in the years following 1920, primarily in Europe and Russia, but to some extent in the United States (see Chapter 4).

Thus it became clear that the change from the vegetative to the reproductive condition in higher plants may often be initiated by some change in the environment. The changing length of day seems to be such an obvious aspect of this environment that it is indeed quite surprising that the discovery was made virtually within our own generation.

3. *Biological Timing*

Perhaps the most impressive thing about the phenomenon of photoperiodism is the implication that the organism is measuring time (see Chapter 8). Thus the cocklebur requires at least $8\frac{1}{2}$ hr of uninterrupted darkness (the so-called critical dark period or critical night) to initiate flowers. Most amazing of all, essentially the same critical dark period is required over at least the temperature range of 15 to 30°C. We can easily visualize time measurement by thinking of the time required for completion of a chemical reaction, but that this should be independent of temperature is not easy to understand.

It is probably safe to say that the formal study of biological timing was first initiated by Garner and Allard in 1920, although previous work was closely related, and man has always had a sense of time and probably suspected that other animals, at least, shared this. In the late 1920's zoologists observed that bees could be trained to feed at certain times of day. About this time rhythmical movements of leaves and other organs were clearly shown to continue under constant environmental conditions. In spite of this early work, the idea of biological timing did not occupy the minds of many biologists until the 1950's.

It is now known that this phenomenon might be a general manifestation of virtually all living things. This cannot be stated as yet with absolute certainty, but nearly all lines of evidence seem to converge on this generalization. As we shall see in Chapter 8, photoperiodism is only one of many examples of biological time measurement. It has become a problem of fundamental importance, and obviously the flowering process is an excellent example.

4. Biochemistry

Following time measurement (critical night), a flowering hormone seems to be synthesized in the cocklebur leaf. Energy and proper substrates are required. As mentioned above, the dark period is ineffective unless it is preceded by a period of high intensity light — a period of photosynthesis. The response to the light environment also is biochemistry, since it is mediated through a pigment system, although we might refer to this process as *photo*-biochemistry (a subdivision, perhaps, of photochemistry).

The light response is a most interesting process from the biochemist's viewpoint, since it apparently involves a trigger type reaction. In photosynthesis the absorbed light energy is converted to chemical bond energy, but in flowering the quantity of light energy involved is extremely small, and rather than itself *causing* flowering to occur, it turns the switch which then influences the biochemistry of the flowering process. The remarkable fact is, that in the case of the cocklebur or other short-day plants, turning the switch during the dark period with light leads to an inhibition of flowering, while in long-day plants this same switch promotes the process. As we shall see in Chapter 7 the pigment system also controls many other phenomena of plant growth.

Action of the flowering hormone at the shoot tips must also be biochemical, as is the very process of growth itself. Certainly biochemistry is the spirit of modern biology. No other approach has contributed so much in recent years. Thus it is somewhat disappointing to learn that virtually nothing is known with certainty about the biochemistry of the flowering process. We have some ideas, and they will be discussed in Chapter 9, but concrete and specific information still belongs to the future.

5. *Morphogenesis or the Origin of Form*

The final aspect of the flowering process which we will consider is the transformation of the meristems from the vegetative to the reproductive condition. In the cocklebur and perhaps most plants which are sensitive to photoperiod, the flowering hormone is translocated from the leaf to the shoot tips, where it causes this redirection of growth. The change seems to begin essentially at the moment when the hormone arrives, and the subsequent rate of development of the flower buds is proportional to the amount of hormone which reaches the meristems.

It could well be that this aspect of the flowering process has the most fundamental biological significance. When we think of the nearly infinite variety of biological structures, the origin of form takes on considerable interest. Here is the real essence of the relationship between diversity and uniformity in biology. Our observations have convinced us that morphogenesis follows essentially the same pattern in all living things: cells divide, enlarge, and then specialize (differentiate). The secret of diversity in the resulting tissues, organs, and organisms must lie in the differentiation step. During growth the cells are specializing in specific ways that will result in special final organized forms or structures. The degree of coordination of this process is truly fantastic. Only cancerous growth and the occasional monster seem to have escaped this coordination. Since morphology is an inherited trait, all of this coordination and final structure is under control of the genes.

In flowering we have an excellent situation for study of this phenomenon. The shoot tip carries out the intricate steps of morphogenesis which produce stem and leaves, with branches and their shoot tips in the leaf axils. Upon arrival of the flowering hormone all of this changes. The complex flower, with a highly specific form for each kind of plant, is now produced. It appears that the genes which ultimately control the production of leaves and elongated stem are turned off, and the genes for flowers are turned on. Since the flower parts may be thought of as modified leaves, it seems likely that only some of the first set of genes are turned off, but obviously some new ones are turned on. And all of this takes place in response to our chemical substance, the flowering hormone. If morphogenesis in general is a response to chemical substances, study of the flowering process from this standpoint becomes of extremely broad and fundamental significance.

The disappointing thing is that we know little more about the topic than has already been stated above. The problem will be mentioned again in the last chapter along with a summary of what has been done so far to try to solve it, but at this stage very little is known.

The initiation of flowers is a change-over from the indeterminate to the determinate form of growth. The indeterminate form of growth of a plant stem confers potential immortality to the vegetative plant. Leaves, stems and branches can be produced indefinitely, so long as the apical meristem remains alive and active. Thus cuttings might well be taken from the 4000-year-old pine trees in the Sierra Mountains of California, and these might grow for another 4000 years, after which other cuttings could be taken, and so on potentially forever.

The determinate form of growth, typical of most animals, leads to death. The embryo grows, essentially in all directions, until maturity is reached, senescence finally sets in, and death ends the process. Preservation of the species depends upon starting over, so to speak, as single cells from male and female are combined to produce the zygote and new individual.

The flower and subsequent fruit also have the determinate growth form. In a sense, the vegetative meristem is "used up" when it develops into the flower. It is no longer capable of producing the plant body as a whole, but only the determinate flower parts — and of course the gametes which may form the zygote and new individual plant. Thus the initiation of sex organs exchanges the potential of immortality for the possibility of combining germ plasm to produce a new individual. Might we thus conclude that sex leads to death?

The flowering plants have solved the problem in various ways. The true annuals have made the sacrifice. If in nature the environment (or their own internal metabolism) causes them to convert all their buds to flowers during their first year of life, then they only live one year, preserving the species until the next year only in the form of the seed. The cocklebur is an excellent example. It will live for years (potentially forever) in a greenhouse with artificial light where it never is exposed to a dark period exceeding $8\frac{1}{2}$ hr. It can be killed within 2 months, however, by exposing it to a number of long dark periods which cause a profuse production of flowering hormone and subsequent conversion of all buds to flowers.

Many biennials require the cold of winter for initiation of flowers. Thus they grow vegetatively the first year, producing many leaves and often storing reserve food in the root or some other organ. Again, if they are kept warm, some of them will grow in this manner indefinitely. But following the first winter they form flowers, use up their buds, and die. So-called winter annuals may germinate in the fall, respond to the low temperatures of winter, and flower and die the next spring soon after producing only a limited vegetative growth.

Another group of plants survives more than one year but finally flowers once and then dies. Other mechanisms, for the most part not very well understood, are involved. The "century" plant may grow vegetatively for a dozen or so years and then send up a flower stalk and die. It may be that formation of the flower stalk depends upon storage of ample food, but other factors could also be involved (accumulation of a flowering hormone which does not reach a sufficiently high level for a decade or so?). Certain bamboo plants may live for up to 50 years, flower once, and then die. Perhaps in such cases flowering does occur in response to the environment, but the plant is not able to respond until it has reached a certain age. We would say that it must reach "ripeness to flower".

The true perennials convert only a certain portion of their buds to flowers, keeping others in the vegetative condition indefinitely. Often a certain age must be reached, but flowering then continues at yearly intervals. The plant may produce flowers only from the axillary buds and never from the primary, central shoot tip, which always remains vegetative. Or the converse may be true.

One of the most striking perennials is the so-called bottle brush tree. Here the flowers form only in the axils just below the apical meristem itself. The central meristem continues to grow, producing axillary flowers (with long stamens, giving the bottle brush appearance) during the flowering season and then finally leaves and stem again.

The familiar woody plants of street, garden and forest are perennials, and yet they are the ones that have been studied the least. This is probably because experimentation with them requires a great deal of time, and the annuals are much easier to work with.

At any rate it should be clear that much remains to be learned about flowering. We will proceed with our survey of what is known.

CHAPTER 2

THE MANY RESPONSE TYPES

A STRIKING feature of the initial work of Garner and Allard was that the principles which they discovered seemed to unify and simplify our understanding of nature. The flowering plants could be classified into one of three simple groups: day-neutral plants, short-day plants, and long-day plants. The simplifying beauty of these discoveries still remains, and the classification system which follows will be based primarily upon that of Garner and Allard. Right now, however, we are discovering that the details of flowering response are very different among the species of flowering plants. Indeed, it now appears possible that no two species will be found to operate exactly alike. For that matter, it is a common thing to find differences in flowering response between varieties within a species or even individuals within a variety. Response to the environment is such a complex thing that only slight differences in genetic make-up seem to influence it.

The three day-length responses of Garner and Allard have become complicated. To begin with it is now known that plants do not necessarily respond in an all-or-none way to day-length. Many plants are known which will flower on any day-length but which flower earlier on short-days. Other plants will flower on any day-length but will flower earlier on long-days. Such a promotion in flowering by a given day-length is referred to as a quantitative response, and such plants may be termed quantitative short-day or long-day plants. Qualitative short-day or long-day plants are those which have an absolute requirement for a specific day-length, so far as their flowering is concerned. We will often refer to the qualitative type of response as absolute, a term phonetically unrelated to quantitative and therefore not easily confused with it.

In addition to this quantitative-absolute complication, plants have been discovered which require neither long-days nor short-days but some day-length intermediate between these two. Thus if the days are either too short or too long, the plant remains vegetative, but

11

when the day-length is exactly right, flowering ensues. Examples of the converse situation are also known.

Another complication was discovered in the early part of the 1930's, but it has been largely ignored until recent years. Thus some plants require a combination of day-lengths: either short-days followed by long-days or long-days followed by short-days. Again it appears that there is an absolute and a quantitative response; some plants require this combination, and others are only promoted by it.

Even before the discovery of the day-length response it was known that some plants were promoted in their flowering by exposure to low temperature. Work since that time has revealed a rather bewildering series of responses of this type. Some plants have an absolute requirement for a low-temperature treatment. Others may flower without the low-temperature treatment but are quantitatively promoted by it. Some respond to lower temperatures by immediately developing flower buds; others produce flowers only after they are returned to warm temperatures. Still others are promoted in their flowering by exposure to warm temperatures or to alternating temperatures.

In addition to the flowering response to a specific day-length or to a specific temperature treatment, there are a number of interactions between day-length and temperature. As one might expect, the measurement of the day-length or the dark period by the plant is influenced by the temperature (but much *less* than one might expect). In addition, there are some rather striking effects, such as replacement of a day-length requirement by a given temperature treatment, or an absolute day-length requirement at one temperature and no day-length requirement at another. Furthermore, any given plant may have virtually any combination of day-length or temperature requirements along with any of the interactions between day-length and temperature.

The Dimensions of Response Type

Thus any classification of the response types becomes a formidable task. A preliminary classification recently appeared in a paper by P. Chouard (13) who is at the Sorbonne in Paris. He considered an absolute requirement for low temperature, a partial requirement (which he divided into "great" and "small"), and no requirement.

These he combined with an absolute requirement for long-days, a partial requirement ("great" or "small"), no requirement for long-days, and a requirement for short-days. Considering the "great" and "small" requirements, there were 20 possible response types, and he cited examples of plants which are known to belong to each of the 20 groups.

Chouard's table of response types omitted a number of categories such as the quantitative short-day plants and the intermediate and combination-requiring plants. There is also no mention in the table of temperature and day-length interactions. Considering only what has been discussed above, it can be seen that any chart purporting to classify the many response types should have at least three dimensions: one for the day-length response, one for temperature response, and one for the interactions. This is the approach followed in the rest of this chapter.

It is very easy, however, to think of other dimensions which could be added to such a chart. We will ignore these because as a rule, they are *known* to be of importance only in isolated cases. Light intensity, for example, is known to influence flowering in an absolute way in certain species; that is, flowering occurs only when the light intensity reaches a certain level (high in some cases, low in others). Light quality is known to be of paramount importance in virtually all of the plants which respond to day-length, and there are certainly many kinds of response to light quality. For example, with *Salvia*, the proper mixture of wavelengths during the day caused a red light interruption of the dark period to promote instead of inhibiting. Some of these responses to light quality are discussed in Chapter 7, but they will not be used in the classification which is attempted here.

In the early days of plant physiology, mineral nutrition was thought to play a very important role in flowering of all plants. It was theorized that the nutrient status of the soil would control the flowering of certain species. Since the work of Garner and Allard, however, it has been shown that the soil seldom exercises any real control on the flowering process. It is true that quantitative responses are easy to demonstrate: high nutrient levels in the soil promote the flowering of cocklebur (see Table 5–1 in Chapter 5), but the cocklebur will not flower at all unless the day-length conditions are right. Low nutrient conditions seem to promote the flowering of other species, and such knowledge has considerable horticultural application. This information could be taken into account in a classification system. In general, however, experimentation has not proceeded far enough to allow us the use of such criteria in our classification.

One species is known which seems to respond in an absolute way to soil conditions. This is the sugar cane, which in Hawaii will flower profusely when it is grown upon one type of soil, and remain absolutely vegetative in a near-by field on a different soil type (45). The same response is obtained when plants are grown on the two soils in pots in the greenhouse.

Still another dimension might be the age of the plant. *Datura* (the jimson weed), for example, is promoted in its flowering by short-days when it is young, but in its old age it flowers equally well on any day-length. Since the age of a plant is much more important in some species than in others, this could also be used as a criterion in classification.

THE APPROACH TO THE CLASSIFICATION SYSTEM

The following classification system is restricted to conditions which might be encountered in nature: combinations of day-lengths obtainable on a 24-hr cycle and temperatures which might be encountered during the normal day and night in nature. The responses to colored lights, interruptions of the dark-period, short interruptions with extreme temperatures, etc., are not used. These will be discussed later when we investigate the mechanism of the flowering process. A classification system for each of the three dimensions listed above is outlined, and some combinations of these responses known to have examples among the flowering plants are listed, along with examples, in an appendix.

Two things were found to be of considerable help in working up the outline. One was a system of symbols which allows one at a glance to know the response type of any given species. The symbols are listed along with the response types in the classification below and in the appendix. Another handy mechanical help was a punched card filing system as illustrated in Fig. 2–1.

Before going to the classification system, one more point should be emphasized again. A short-day response is not simply a response to "short" days. The response (on 24-hr daily cycles) is to day-lengths which are *shorter* than a maximum or night lengths which are *longer*

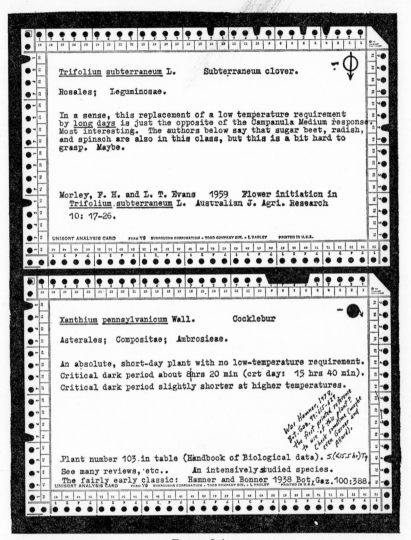

FIGURE 2–1

Two cards used in working out the response types. Notches around the
edge are punched according to a code which indicates the spelling of the
genus name at the top, the day-length and temperature responses on the
left, and the interactions on the right. The bottom remains unpunched
for future complications! When all of the cards are placed in a stack, a
needle may be run through the hole which represents the response one
might be seeking, and when the cards are shaken, those which have been
punched at that number fall out.

than a minimum. Thus knowing that a plant flowers when days are "short" or "long" does not indicate its response type. Both cocklebur and henbane flower on 14-hr days, but cocklebur is a short-day plant while henbane is a long-day plant. Cocklebur flowers when the days are *shorter* than about 15½ hr, and henbane flowers when the days are *longer* than about 11 hr.

OUTLINE OF THE THREE DIMENSIONS

Since the temperature response precedes the day-length response as a rule, there would be logical grounds for listing the temperature response first in the following classification. The system would thus be more natural. Yet I have arbitrarily decided to base the system on the day-length response, primarily because this was the first classification of response as established by Garner and Allard, and also because the day-length response is emphasized in this book. Furthermore, the temperature response is a bit difficult to use in a classification because one should really distinguish between delayed and direct temperature effects on flowering, and this could nearly double the categories. The present system, then, like Chouard's, is far from complete. It takes into account more kinds of response, but a number of important responses, which are probably physiologically quite different, are grouped together. A truly comprehensive classification will have to await many more studies, probably using controlled conditions.

It has been suggested that multiplication of the response types can only lead to confusion. This may be true, but any search for truth seems to lead to such a division and multiplication of types. It is merely a description of our present understanding of nature. The impressive thing is that the few basic types of Garner and Allard can still provide a framework for the many details which keep coming to our attention.

I. PHOTOPERIODISM: THE FLOWERING RESPONSE TO DAY-LENGTH
   (DURATION OF PHOTOPERIOD OR DARK-PERIOD)

- 1. Day-neutral plants. No flowering response to day-length.
  2. Flowering response to a single day-length.
     A. Quantitative response to day-length (promotes, but not essential).

◉ (a) Quantitative short-day plants.

○ (b) Quantitative long-day plants.

B. Absolute or qualitative response to day-length (specific day-length required for flowering).

● (a) Short-day plants.

○ (b) Long-day plants.

3. Combination of day-lengths or special day-length required for flowering.

A. Quantitative response.

◒ (a) Quantitative intermediate - day plants. Plants flower under all day-lengths but flower best when the day is neither too short nor too long.

◖ (b) Quantitative short-long day plants. Plants flower under all day-lengths but flower best when they are given short days followed by long days.

◗ (c) Quantitative long-short day plants. Plants flower best when given long days followed by short days.

B. Absolute or Qualitative response. Plants flower only when they receive a special day-length or combination of day-lengths.

⬭ (a) Intermediate-day plants.

◐◖◖ (b) Short-long-day plants.

◑◗◗ (c) Long-short-day plants.

II. Effects of Temperature on Flowering

— 1. No effect of temperature on flowering.

2. Quantitative response. There is no absolute requirement for a specific temperature, but such treatment promotes flowering. This could be subdivided according to whether the response is inductive (delayed) or non-inductive (direct). That is, the particular temperature treatment may not cause any visible change in the plant at the time of the treatment, but subsequently the plant will flower. Such a response would be considered delayed or inductive. On the other hand, plants are known (e.g. Brussels sprouts) in which the formation of

flower buds begins to occur as soon as the proper temperature is applied, and in some of these cases if the proper temperature conditions are removed, the development of the flower will cease. Such a response is referred to as being direct or non-inductive. We will arbitrarily restrict the term vernalization to delayed (inductive) promotion of flowering by low temperature (but see discussion in Chapter 4).

↕   A. Low temperatures promote flowering. In known examples, this is usually, but not always, an inductive response.

↑   B. High temperatures promote flowering. This is probably always a direct response.

↕   C. Alternation of temperature promotes flowering. This is also a direct response.

3. Absolute or qualitative response to temperature. Flowering in such plants is absolutely dependent upon a particular temperature treatment.

↓   A. Low temperature is required for flowering. Called vernalization if the response is delayed (inductive).

↑   B. High temperature is required for flowering. Usually a direct response.

↕   C. Plants flower only if temperatures are alternated. Also direct.

III. Interactions between Temperature and Photoperiodism

1. No qualitative change of day-length response type due to change in temperature.

A. No known interaction with temperature. Of course, all day-neutral plants would fit in this category, but placement of other plants in this category usually implies only a lack of study. It seems quite likely that any such photoperiodically sensitive plant which is studied in sufficient detail would prove to have some sort of interaction between day-length and temperature. It should at least fit in one of the next two categories. The majority of plants listed in the appendix had to be placed in this category, but future work is bound to change this situation.

○ B. Increasing temperature during the day-length treatment increases the length of the critical dark period. The critical dark period may be defined as the minimum time length in darkness which will promote the flowering of short-day plants or inhibit the flowering of long-day plants.

○ C. Increase in temperature decreases the length of the critical dark period. Probably the majority of day-length responding plants would fit in this category if sufficient work had been done to permit their classification. Placement in the last category (III.1.B) is very rare, and on theoretical grounds it is not to be expected.

2. Temperature influences day-length response type when applied during the day-length treatment.

A. Most qualitative or absolute requirement at low temperature.

(a) High temperature: quantitative. This means that a plant has an absolute requirement for either short- or long-days at relatively low temperatures, but at high temperatures short- or long-days only promote flowering but are not essential.

(b) High temperature: day-neutral. The plant may either require or be promoted by either short-days or long-days at low temperatures, but at high temperature it is day-neutral.

(c) High temperature: response changes. Such a plant might require short-days at low temperature and require or be promoted by long-days at high temperature. This category could, of course, be easily subdivided on the basis of quantitative and qualitative response. It is assumed that the plant might be day-neutral at the intermediate temperature, although an intermediate day requirement cannot be eliminated. Of course this category is very interesting, with only limited examples.

B. Most qualitative or absolute requirement at high temperature.

  ⊙-●   (a) Low temperature: quantitative.
  ○○

  •-⊙   (b) Low temperature: day-neutral.
  •-●
  •○
  •○

  ○-●   (c) Low temperature: response changes. If the
  ⊙-○   response is absolute at both high and low temperature, this category would be the same as III.1.A.c above, but it is conceivable that a plant might have an absolute requirement for short-days (or long-days) at high temperature and a qualitative response to long-days (or short-days) at low temperature.

3. A change in day-length response type due to temperature treatment.

  e.g. ●   A. Day-length and temperature treatment are interchangeable; that is, the plant will flower in response to one or the other. In the known cases low temperatures are required for this effect. Thus we have an interchangeability between photoperiodism and vernalization. If flowering occurs while plants are in the low temperature conditions, this becomes category III.2.B.b above. Here there is a day-length requirement unless the plants are made day-neutral with a low temperature treatment.

  e.g. •↓●   B. Temperature treatment inductively changes the day-length response type. In the very few examples at hand this is again a low-temperature response. Care must be taken here to separate this response from the ones listed under III.2 above. For example, if a plant requires short-days at low temperature, but at high temperature it has no day-length requirement, it would be placed in III.2.A.b above and not in this category. This category is reserved for the inductive response in which plants *not* given low temperature may be day-neutral, but *following* treatment with low temperature, long-days (or short-days) are required for flowering.

AGE AT TIME OF RESPONSE AND FLOWERING TIME

In addition to the above dimensions it is convenient to know the condition of the plant at the time when it responds to the environmental stimulus and also to know the time when the plant normally flowers in nature, that is, whether it is an annual, a biennial, or a perennial. These factors will not be taken into consideration in the classification, although they are of considerable importance especially in the case of vernalization. The following lists indicate these factors along with the symbols which are used:

Plant Type: (Condition at sensitive time.)

⇂ 1. Seed or young seedling.

↓ 2. Rosette plants (leaves all arising from the root, such as a dandelion), including grasses.

3. Caulescent plants (leaves from a stem).

   ↓ A. Herbaceous.

   ↨ B. Woody (bushes or trees).

Flowering Time:

1. Annuals.

   ↓ A. Winter.

   ↓ B. Spring.

   ↓ C. Summer.

↓ 2. Biennials.

↧ 3. Perennials.

THE CATEGORIES OF RESPONSE TYPE

Having listed the responses to photoperiod, temperature, and the interactions, it should now be possible to work out the theoretical combinations of these factors and discover how many response types might be expected if all combinations occur in nature. This can be done by use of a three-dimensional approach such as that shown in

D

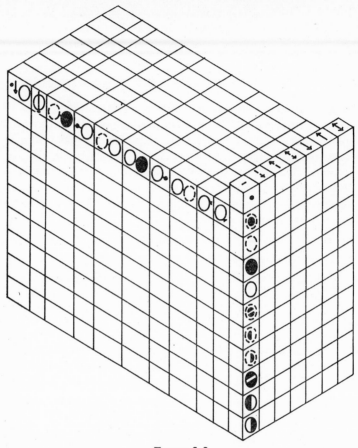

FIGURE 2–2

The possible combinations of response type. There are 777 visible and other solid blocks, each representing a possible combination of response types. Obviously some would not be expected to occur, and other possible categories are not shown, so the number is only an indication of possible magnitude.

Fig. 2–2. Temperature and day-length interactions are, of course, impossible in day-neutral plants since they have no day-length response, and hence this is omitted from the figure. All other categories would be accounted for, even though we might consider them as being rather unlikely. Multiplying by the number of categories horizontally, vertically, and in the third dimension, we see that there are 777 possible plant response types illustrated by the figure. Interactions which are really different response types are classified together in the present system (long-day at high temperature, day-neutral at low falls in the same category as quantitative long-day at high temperature and day-neutral at low). Thus it would be quite simple to increase the response types beyond 777, but we will leave further complications to the future.

The classification in the appendix includes mostly plants in the rather readily available literature, mostly the review by Chouard (13) and a list in *The Handbook of Biological Data*.[1] Some plants were left off because of lack of information about their response; e.g. the response to cold may be stated, but response to day-length may not. Of those plants which are included, it is quite likely that a majority will prove to be in a wrong category as experimental work progresses. This is most true of the temperature and day-length interaction categories. Very little information was available about these categories, although it is quite likely that most plants would show such an interaction if it were studied. Main changes in the future should be movement of a plant from a no-interaction category to a specific interaction category. The Japanese morning glory is an excellent case in point. During the 1950's it has been studied so intensively by the Japanese that information about it probably equals information about any other plant, and yet it was only in 1961 that its day-neutral response at low temperature was demonstrated. The response is not striking but quite real. Previously this plant was considered to be a typical short-day plant with temperature interactions only in so far as the critical dark period was concerned. Nearly all of the plants in the appendix have been studied only very superficially compared to the Japanese morning glory. The present policy was to place the plant in a "no response" category if in doubt.

[1] The following reviews also contain lists of plants, showing response type: references 17, 24, 25.

CHAPTER 3

ECOLOGY AND THE FLOWERING PROCESS

THE time of flowering is a response to environment; it is an interaction of the plant with its environment. Thus flowering is primarily an ecological phenomenon. Yet comparatively little study of the flowering process has been made from the purely ecological standpoint. The mechanisms going on within the plant have proven to be so fascinating that nearly everyone interested in the subject has become preoccupied with this phase of the study. Indeed the bulk of this book will be concerned with the physiological mechanisms of the flowering process. Nevertheless, to try to set the proper perspective, we will devote this chapter to the ecological aspects of flowering: the flowering process in natural habitats as a response to environment.

The environment itself is a fit subject for extensive study. There are climatologists and others who are primarily concerned with temperature, humidity, wind, soil, and light as they occur and vary in nature. We do not have space to investigate these topics in detail, but we will review seasonal changes in temperature and light, since floral initiation is most commonly a response of the plant to these factors.

NATURAL TEMPERATURES AND FLOWERING

Of course, temperatures are extremely variable and dependent upon latitude, elevation, time of year, and various other factors. It is easy to generalize, however, that temperatures in the tropics are seldom low and that winters become colder as one moves towards the poles or higher in elevation. Few qualitative plant responses to slight temperature changes are known, and it seems quite appropriate that important steps in the plant's life cycle should not be carefully tuned to slight changes in temperature, since this factor in the environment is so variable at a given location from year to year. Many plants do require the low temperatures of a cold winter for flower initiation. Thus response occurs if there is a cold winter as contrasted with a winter in which temperatures never even drop to the freezing point. If a plant will not flower unless it has experienced a number of days (a few weeks to a few months) of near freezing temperatures, then it is protected from blooming in the fall, when sudden frosts might stop growth of the flower.

It is clear that response to the cold temperatures of winter might be of strong ecological advantage, but flowering in response to 18° instead of 22°C would seem to be of little advantage to most plants. Yet when our researches become sufficiently detailed we may discover examples in which response to such slight changes is of ecological significance. This could be the case with certain alpine plants, where maximum day temperatures are commonly below 20°C and seldom above. Most of our present knowledge, however, is concerned with a response to near freezing temperatures extended over a considerable period of time. We will consider this in the next chapter and reserve the rest of this chapter for discussion of light and the flowering process from the ecological standpoint.

FLOWERING IN RESPONSE TO NATURAL LIGHT (36)

1. *Light as a Natural Variable*

Light varies in three ways: quality (color or wavelength), intensity, and duration. We might also think of light in a quantity or total energy sense, considering both intensity and duration. High intensity light applied for a short interval of time supplies a quantity of energy which is only equalled at lower intensity by a long period of time. All of these factors vary more or less continuously in nature and all are important to the photoperiodism aspect of the flowering process.

2. *Natural Light Qualities and Flowering*

Perhaps quality of light varies the least in nature. Figure 3–1 illustrates the quality of sunlight, showing the intensity (relative energy) at any given wavelength or color twice during the day. Note that about half of the area under the curve falls within the region to which our eyes are sensitive — the so-called visible light. Nearly half, however, consists of longer wavelengths, the infra-red.

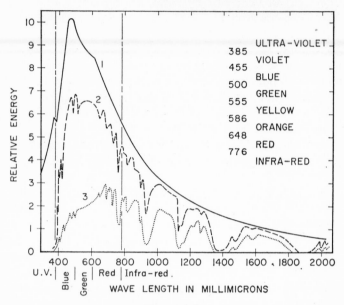

ULTRA-VIOLET
VIOLET
BLUE
GREEN
YELLOW
ORANGE
RED
INFRA-RED

FIGURE 3–1

The solar spectrum. Curve 1 shows the normal solar energy distribution of radiation estimated for outside the atmosphere. Curve 2 shows the same at the surface of the earth on an average day in Washington, D.C., with 1.37 cm of precipitable water in the atmosphere and the sun 25 degrees from the zenith. Curve 3 shows the shift towards the red when the sun is 78.7 degrees from the zenith (altitude of 11.3 degrees), producing an air mass about 5 times that of Curve 2. Other conditions about the same as for Curve 2. Redrawn from Otis F. Curtis and Daniel G. Clark, 1950, *An Introduction to Plant Physiology*, McGraw-Hill, New York.

The energy of this infra-red light warms our planet, although so far as we know, organisms do not respond to it photochemically. It is probably a little more efficient in warming the earth than the visible light, because infra-red is absorbed more efficiently by most things on the surface of the earth. Water, for example, absorbs much of the infra-red, while the visible is mostly either transmitted or reflected. Carbon dioxide also absorbs infra-red light but not visible light.

Note that the peak of the midday curve falls approximately in the region of green light. Our eyes are most sensitive to green light, but most plant processes are relatively insensitive to light of these wavelengths. Quality changes that occur in nature are usually considered to be of minor importance. The amount of moisture or minute particles in the air such as smoke or smog may have an effect on light quality. Ultra-violet light is absorbed by the atmosphere (mostly the ionosphere) and consequently as one goes higher in elevation above sea level, a slightly higher proportion of ultra-violet light is encountered. In general, red wavelengths penetrate these atmospheric constituents more readily than shorter wavelengths (for example, blue). Thus when the sun is very low on the horizon in the morning or the evening, and its rays must penetrate a much thicker layer of atmosphere, those reaching the earth may be predominantly red (curve 3, Fig. 3–1). At temperate latitudes, quality also changes slightly throughout the year, since the rays must pass through more air in winter when the sun appears lower in the sky. In the oceans or other bodies of water, light quality changes with depth below the surface. As a rule, the natural changes in light quality are not thought to be very important to the response of living organisms, but interesting examples are being discovered, and detailed study may reveal unexpected phenomena. For example, we now know that the response to red or far-red light (see Chapter 7) might explain the growth habits of plants on the forest floor; thus filtering by chlorophyll of the tree leaves may strongly increase the proportion of far-red light reaching the plants below. Leaf size, stem length, etc., of these plants may be profoundly influenced.

3. *Natural Light Intensities and Flowering*

The response of living organisms to different light intensities is a fascinating subject for study. Figure 3–2 summarizes a number of these responses. Note that essentially no response is capable of utilizing the maximum intensity of full sunlight. Photosynthesis responds to high intensities, but individual leaves are often photosynthesizing at their maximum rate when sunlight is only about one-fifth of maximum intensity, although a complete plant community may respond photosynthetically to higher intensities. On the other end of the scale, it is most interesting to see that some plants are sensitive to intensities much lower than those which can be detected by the human eye. The situation becomes involved, however, since the human eye must respond within a fraction of a second, while plant responses are measured after exposure for a long period of time. Nevertheless, the plant must be "seeing" these extremely low

RANGE OF LIGHT INTENSITIES
FOR BIOLOGICAL RESPONSES

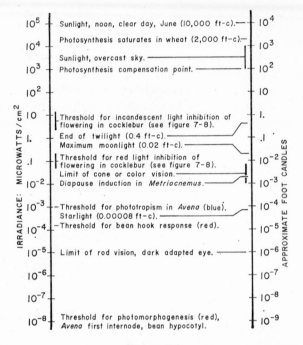

FIGURE 3–2

The range of light intensities for biological responses. The agreement between the foot candle scale and the absolute energy scale (microwatts/cm²) is at best an approximation. The foot candle (ft-c) is not a good unit for measurement of light intensities which relate to any biological response except human vision, since the unit is based upon the ability of the human eye to respond to light. It is an excellent unit for measurement of illumination for reading, working, etc., but its use in other biological research should be avoided. Adequate light measurement must express absolute energy in terms of the wavelengths being measured. Such measurement is still very difficult, but the nature of the light source can be expressed and the intensity given in energy terms. Unfortunately data for some of the responses listed gave light intensity only in terms of foot-candles, as indicated by the lines. Modified from R. B. Withrow (36).

intensities during any given second, even though it must "add up" the response over a long period of time before actual changes in the organism can be observed. The plant response is quite similar to the use of photography in astronomy. The astronomer is able to "see" many objects by allowing their light to fall upon a photographic plate for a number of hours, although his eye, with its short time limitation, would be unable to detect these objects.

It is important to note that response in photoperiodism will occur under very low light intensities — only slightly more intense than those detected instantaneously by the human eye. The time factor is extremely important here again, and we will consider it in some detail in Chapter 7.

Figure 3–3 illustrates the fact that response in many light sensitive plant processes is essentially linear to the logarithm of light intensity

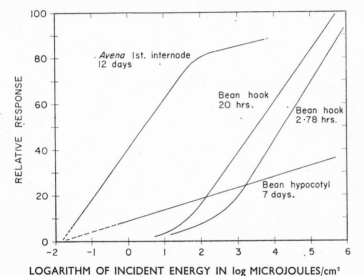

LOGARITHM OF INCIDENT ENERGY IN log MICROJOULES/cm²

FIGURE 3–3

Response to the logarithm of light energy in three stem-growth manifestations. Indicated days and hours are exposure times. Note saturation
in *Avena* curve. Redrawn from R. B. Withrow (36).

rather than to actual intensity itself. Thus the difference in response between quantities of 0.01 and 0.1 microjoules/cm² may be about the same as the difference in response between 0.1 and 1.0 microjoules/cm². In short, the plant is proportionately less sensitive in a given response to very high intensities than it is to very low intensities.

In Chapter 7 we will be concerned with the idea of saturation intensity. Allowing an exposure time of, say, 5 min, one finds that increasing intensity causes an increased response up to a certain point, and then the response stays the same (maximum) even though intensity increases (see *Avena* curve in Fig. 3–3). The intensity level beyond which no change in response occurs is called the saturation intensity. In flowering we are concerned with the light absorbing reaction which informs the plant by adjusting its metabolism, as to whether it is day or night. This reaction may be studied by interrupting with light the dark period which promotes flowering of short-day plants or inhibits flowering of long-day plants. If we allow a reasonable exposure time such as 1 to 5 min, the saturation intensities for these reactions are relatively low, especially compared to sunlight. Even the intensities under rather heavy cloud cover (as low as 100 ft-c) will cause saturation in time intervals of less than a minute (saturation is approached in about one minute at 1% sunlight). Thus the plant, during the day, "sees" much more light than it needs to completely saturate its response capacity in flowering. During the day, then, changes in light intensity due to clouds, are probably of no importance to this part of the plant's life, but the question immediately arises, at what time during dawn or dusk is the plant responding as though it were day, and at what time is it responding as though it were night?

4. *Twilight and the Flowering Response to Day-length*

In experimental work, it is convenient to turn the lights on or off so that the plant changes in a small fraction of a second from conditions of saturating light to complete darkness or vice versa. Obviously in nature the change from light to dark or back to light is not so abrupt. The duration of twilight on clear days will depend upon the latitude and the time of year, since at a given time the elevation of the sun in relation to the horizon and the air mass through which the rays pass will depend upon these factors. In addition many local variables might influence twilight, such as elevation, nearby mountains, atmospheric conditions, and of course weather. Thus the clear sky curves shown in Fig. 3–4 are only representative. They should be improved by new measurements to show the intensities of light (using energy terms instead of foot candles) in the specific wavelength regions known to be of importance in photoperiodism. In Fig. 3–4 intensities are shown on an absolute and on a logarithmic scale as functions of true elevation of the sun in relation to the

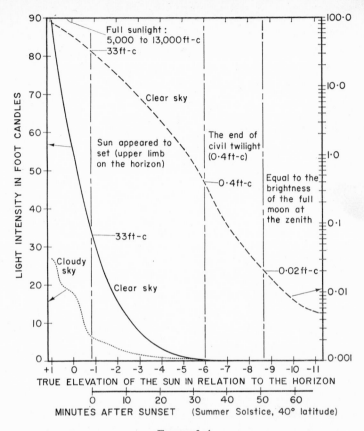

FIGURE 3–4

Light intensity during twilight as a function of the sun's elevation in relation to the true horizon. Clear sky curves were redrawn according to the curves of H. H. Kimball, 1938, *Monthly Weather Review* 66: 279–286. Kimball's curves were drawn by simple estimation through points taken on a number of occasions during 1913 and 1916, and as might be expected there is considerable scatter in the points. Thus the curves are at best only examples of what might occur on any given clear day. The duration of twilight will depend upon latitude and time of year, and the times shown are for 40° latitude on June 21, the summer solstice. The time shown as 50 min would be about 40 min on the spring or fall equinoxes and 45 min on the winter solstice. At 30° latitude on the summer solstice the point indicated as 50 min would be about 41 min, and at 50° latitude the point would be about 69 min. The cloudy day curve is a typical one taken from measurements of A. Takimoto and K. Ikeda, 1960, *Bot. Mag. Tokyo* 73, 175–181. Their low intensity data are not accurate enough to allow plotting on the logarithmic scale, but near the end of civil twilight light intensities are about the same on clear and cloudy days.

horizon. Approximate times are shown for 40° latitude. A typical curve for a cloudy day is also indicated.

As further background material for the discussion of flowering response to twilight, we need some more of the information which is to be discussed in Chapter 7. Plants respond to light or its absence through a reversible pigment system called phytochrome. The situation may be simplified for this discussion as follows: when the pigment system is illuminated with natural light, it is driven in one direction; at the same time metabolic processes are tending to convert it back to the first condition. We can enter light in the following equation by thinking of it in a quantitative sense, a given number of light quanta[2] reacting with a given number of pigment molecules:

n. Einstein's red light +
        m.moles R-phytochrome ⟶ m.moles F-phytochrome

Metabolically

Thus in a given interval of time, illumination with a given intensity will produce a certain amount of F-phytochrome. In the same interval of time another amount of F-phytochrome will be reconverted to R-phytochrome by metabolic processes (we will see in Chapter 7 that far-red light also converts F-phytochrome to R-phytochrome, but this can be ignored for the moment). Thus the amount of F-phytochrome present at any time will reflect the rates of the conversions brought about in one direction by light and in the other direction by metabolism. When the amount of F-phytochrome is great the plant "knows" it is in the light (its biochemistry is adjusted to the light status by the presence of F-phytochrome); when the amount of F-phytochrome is small the plant "knows" it is in the dark.

In the evening, then, as light intensity decreases, a point will be reached at which metabolic removal of F-phytochrome begins to exceed its production by light, after which F-phytochrome will gradually decrease. The rate of decrease after this point will depend upon how fast light intensity is decreasing: if the lights are turned off

---

[2] Avogadro's number (the number of molecules in a gram molecular weight of an element or compound; 6.023×10^{23} molecules per mole) of light quanta or photons is called an Einstein.

instantly, the rate of decrease of F-phytochrome will be rapid; in a typical twilight it might be less rapid. When the level is low enough the dark processes will begin in the leaf, and thus the plant will "know" it is in the dark. Even this could be gradual: the processes beginning gradually as F-phytochrome gradually decreases. Obviously two factors will determine the time when the dark phase of flowering will begin: first, the quantity of F-phytochrome as a function of light and metabolism; and second the sensitivity of the flowering process to F-phytochrome. Plants might well differ in this sensitivity. Some could initiate the dark phases of the flowering process when the level of F-phytochrome has decreased by a relatively small amount, and others might require almost complete removal of F-phytochrome for initiation of the dark processes. As mentioned above (Fig. 3–3) these responses to phytochrome often exhibit a logarithmic relationship to light intensity.

In the morning the situation must be exactly reversed. As light gradually builds up in intensity, production of F-phytochrome will gradually increase until light formation is greater than metabolic removal, and a net amount of F-phytochrome begins to accumulate. When the level is high enough, the dark phases of the flowering process will be stopped. Of course, it is quite possible that the biochemistry of stopping is quite different from that of starting.

It appears from the above discussion that there are two times that could be noted during twilight: the time at which the net amount of F-phytochrome begins to change significantly in response to change in light intensity, and the time at which the quantities of F-phyto-chrome become critical in the flowering process. In theory we could state the first time exactly, but the second time could extend over the period during which the processes of flowering gradually stop or start in response to gradual changes in quantity of F-phytochrome. So far the state of our science has not advanced far enough to study either of these times according to the theoretical treatment given above.

A more practical experimental approach has been to transfer plants to complete darkness at various times during evening twilight, and thus to learn how dim the light intensity must be before plants left in the twilight flower as much as plants placed in total darkness. Morning twilight may be studied by reversing the procedure; plants which were all placed in darkness at the same time the previous evening may be removed from the darkness and placed in morning twilight at various times. Atsushi Takimoto and Katsuhiko Ikeda (74) in Japan have performed a number of experiments of this type. With the Japanese morning glory they found that plants left in the twilight after it had dropped to an intensity of about 10 to 20 ft-c flowered as well as plants which were placed in complete darkness at this time. In the morning, however, plants were inhibited in their flowering if they were moved from darkness to twilight any time after the light intensity had reached 0.1 ft-c. Thus they concluded that the biological night for this plant begins in the evening at about the beginning of astronomical sunset and ends in the morning at the time of the beginning of civil twilight. The principle was applied to other plants, and it was found that sensitivity in the morning or the evening varies considerably from species to species. Thus *Oryza sativa* is relatively insensitive both times, and the biological day-length about corresponds to astronomical day-length. Soybean and Perilla acted somewhat like Japanese morning glory, but cocklebur was very sensitive in the evening (0.1 to 1.0 ft-c) and somewhat less so in the morning (1.0 to 5.0 ft-c).

The effect of clouds during twilight could be quite complicated. If it is merely a matter of reaching a certain critical light intensity, then we could easily state the nature of the complication. Thus in Japanese morning glory the 20 ft-c level might be reached up to 30 min sooner on the evening of a cloudy day, but the 0.1 ft-c level is probably reached at about the same time (5 to 10 min later when cloudy, perhaps) on either a cloudy or a clear morning. The problem is, it is not the time of critical intensity which we need to know, but the time of the critical amount of F-phytochrome. This problem has yet to be studied, but it should be fairly apparent that clouds during sunrise or sunset could indeed influence flowering to a certain extent, and this might readily account for the observation that many plants sensitive to photoperiod do not always flower at *exactly* the same time each year. Of course, plants that require many cycles for induction would tend to average out the weather anyway.

An interesting sidelight of the Japanese work was the finding that Japanese morning glory, although it is an absolute short-day plant, is induced to flower in the open in Japan near the end of June when the days are longest. Its critical dark period is shorter than the shortest night at that latitude. It apparently fails to flower in response to the longer nights preceding the summer solstice because temperatures are too low. When it does respond, about the time of the longest day of the year, it is induced only minimally, and thus only the lateral buds produce flowers, while the terminal one continues to grow vegetatively for quite some time. If the terminal bud forms a flower (as finally happens when the nights become somewhat longer) then growth ceases At least it should be quite clear that short-day plants do not always need short days in the usual sense of the word if they are to become reproductive.

5. *Natural Photoperiods and Flowering*

The time at which the sun will cross the horizon at sunrise or at sunset can be predicted for any location with the high precision of astronomy. Figure 3–5 shows the relationship between day-length and time of year at four different latitudes. This is, so to speak, a relationship that the plant can count on. Surely there would be no better way in nature to ascertain the time of the year than to measure the length of day at a given latitude. Of course, as indicated above, complications arise from the duration of twilight and dusk, especially as these are influenced by clouds, mountains, etc., but because of the low intensities required for response, the complications are probably rather minor. As shown in Fig. 3–4, clouds do not influence twilight intensities much when it is nearly dark. The same is true of mountains or other obstructions.

If time of year is ascertained by measurement of day-length, the precision with which length of day is measured will determine the accuracy of measurement of season. This accuracy is strongly influenced by the time of year, because the difference in length of day between any two successive days is not constant throughout the year, but varies according to the curve shown in Fig. 3–6, which may be derived directly from Fig. 3–5. Figure 3–6 shows that the rate of change in day-length is least[3] near the summer and winter solstices when the days are approaching their longest and their shortest durations. For more than half the year, however, from February to the middle of May and from August to the first part of November, the rate of change in day-length is very high and very constant (about 2.6 min per day at 40° latitude). Thus a plant must measure time

[3] Near the solstices the rate of change (the acceleration and deceleration) is very great, but we have no evidence that plants might detect this.

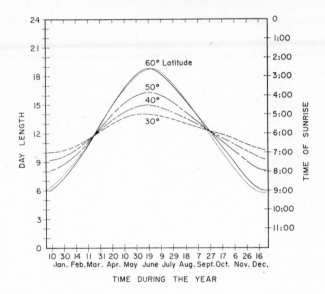

FIGURE 3-5

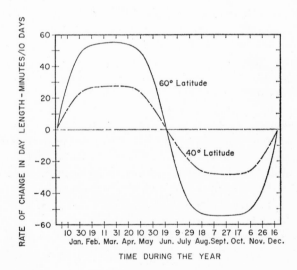

FIGURE 3-6

somewhat more accurately in summer (or winter) than during the spring and fall months. For example, during spring or fall a plant at 40° latitude must measure the diurnal cycle within about 26 min if it is to flower within a 10-day period encompassing the "proper" time. If the night length is measured, 26 min is about 4.3% of a 10-hr night coming at the beginning of August or 3.6% of a 12-hr night near the end of September. If, however, the response occurs about July 9, the 9-hr night will have to be measured with an accuracy of 10 min to insure flowering within the 10-day period, requiring an accuracy of about 2%.

Another problem which arises in trying to understand the responses of plants in their natural situations by using data obtained in our experimental studies, is that we have yet to study in any detail the ability of a plant to measure time as the day-length is changing. It is clear, however, that many plants measure time more accurately when they receive a number of repeated cycles than when they receive only one or a few. With cocklebur this is evident when the effects of only 5 cycles are compared with those of a single cycle, as in Fig. 3–7. With a single cycle the plants of a population varied as much as an hour among themselves in their measurement of the critical night. With five cycles this error is reduced to about 20 min.

Response to day-length implies directly a measurement of time.

FIGURE 3–5

Day-length and time of sunrise at various latitudes as a function of time during the year. Only the dotted curve shows true day-length (60° latitude), the others show sunrise times. The difference is slight and can probably be ignored. The latitudes shown pass near the following geographic locations:

    30° latitude, New Orleans, Cairo;
    40° latitude, Denver, Philadelphia, Madrid, the "heel" of Italy;
    50° latitude, Vancouver, Winnipeg, the southern tip of England, Frankfurt;
    60° latitude, Anchorage, southern tip of Greenland, Oslo, Helsinki.

Data from *Astronomisch-Geodätisches Jahrbuch*, 1954. G. Braun, Karlsruhe.

FIGURE 3–6

Rate of change in day-length at two latitudes as a function of time during the year. Data from *Astronomisch-Geodätisches Jahrbuch*, 1954, G. Braun, Karlsruhe.

E

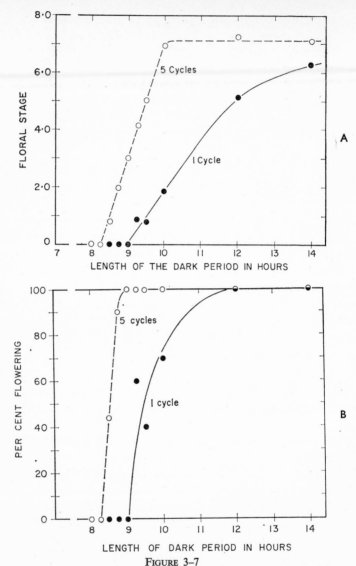

FIGURE 3–7

The flowering response of cocklebur plants exposed to various dark period lengths, applied once or five times. A (top) shows response plotted according to the system of Floral Stages described in Chapter 5. B (below) shows response plotted as percentage of plants within a treatment which flowered. The experiment was performed by the advanced plant physiology class at Colorado State University beginning May 14, 1962. Plants examined after 8 days. Data previously unpublished.

In Chapter 8 we will discuss the problems connected with time measurement, but as a preview it can be stated that it is very difficult to imagine any mechanism which will account for an accurate time measurement, yet be relatively insensitive to changes in temperature. No one has yet proposed a scheme involving known physical processes of plants such as diffusion or osmosis, which will adequately account for time measurement. If it is a chemical process it must be temperature compensated in some very interesting way. Changes in temperature of 10°C are not at all uncommon in nature, yet such a change will typically change the rate of a chemical reaction by a factor of 2. Since temperature often varies this much at a given season from year to year, this would obviously completely destroy any sort of season detection by measurement of day. We can, then, look for some sort of temperature-insensitive, time-measuring process in the plants which are sensitive to length of day or night.

The Many Response Types in Nature

1. *Response Type and Plant Distribution*

In the last chapter, many different plant response types were outlined, based upon environmental changes in light and temperature. How would we expect these to be related to plants in natural habitats? Obviously as one goes farther north the days in the middle of the summer are longer, and one might expect to find more long-day plants. Indeed plants which require long-days for flowering are more common in the north. At the equator, length of day is not changed throughout the year, and day-neutral plants should be common. Short-day plants should be those which bloom in the spring or late summer or fall, most likely in temperate climates where early spring and fall are not too cold. Such is the case. Garner and Allard made such observations in the years just following their initial discovery.

As pointed out earlier, however, the actual day-length is not the important criterion in determination of day-length response type. The criterion is rather, whether the plant flowers in response to increasing or to decreasing day (or night) lengths. Thus short-day plants, for example, can and do flower at suitable seasons in either the tropics or the far north.

There is often a very close relationship between response to day-length and latitudinal location of a given plant. H. A. Mooney (66), a student of the ecologist Dwight Billings at Duke University, has studied the day-length requirements of a small alpine plant, *Oxyria digyna* which grows not only in the alpine tundra of mountain ranges in North America, but also in the Arctic. He found that samples collected from various latitudes throughout the Rocky Mountains and the Arctic each responded in a specific way, depending on the location from which it was collected. As collections were made farther and farther north, longer days were required to induce flowering. Even though all of his samples belong to the same taxonomic species, they were genetically different and very accurately adapted to the region in which they were found.

In recent years, other studies of this type have been made with increasing frequency. Workers at the North Carolina Experiment Station (62), for example, have studied more than 30 species and varieties of cotton (*Gossypium*) under a combination of day-lengths and temperatures. They found a wide diversity within the genus. Most were favored by short days and cool nights, but at least one species clearly initiated flowers more readily under long-day conditions.

Perhaps the most detailed work of this type with a single genus has been done by Bruce G. Cumming (47), in the Canada Department of Agriculture at Ottawa. For a number of years now he has been studying the responses of pigweed (*Chenopodium*) to environmental conditions, his work being at least partially directed towards finding better ways to control these weeds in agriculture. He has determined the sensitivity of 33 species to day-length, for example. He found 25 short-day plants and 8 long-day plants (at least in a quantitative sense). With some of these, germination, response to nutrients, chromosome numbers, lengths of shoots, primary stems, and petioles, and number and sizes of leaves as well as flowering have all been studied in various environments. In the course of this work Cumming discovered the special variety of *Chenopodium rubrum* which will respond to a single dark period and flower in a petri dish as a seedling. Figure 3–8 shows the flowering response of six varieties of *Chenopodium album*, collected from different latitudes in North America. Some (varieties E, D, and F) are quite clearly short-day plants, but

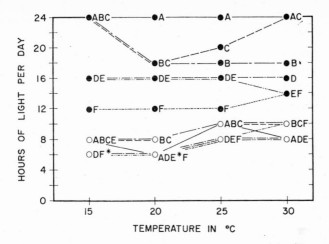

FIGURE 3–8

Flowering response to different day-lengths (24-hr cycles) of six varieties of pigweed (*Chenopodium album*), collected from various locations in North America as follows:

A 62° 46′ N; Yukon	D 50° 10′ N; Saskatchewan
B 60° 52′ N; Yukon	E 49° 58′ N; Manitoba
C 60° 47′ N; Yukon	F 34° 20′ N; California

For a given variety, flowering occurred at a given temperature on the two day-lengths indicated for the variety and at all day-lengths between the two indicated day-lengths. All even numbered day-lengths were tested. Except for the two points marked with an asterisk, all plants died on day-lengths shorter than those indicated by the lower curves. Some sets of plants on day-lengths shorter than those marked with the asterisk remained alive but vegetative. There was considerable variation among treatments in the quantitative aspects of the flowering response. This could not be shown in the figure. Curves drawn from data supplied by Bruce G. Cumming.

at 30°C variety A acts as a quantitative long-day plant (not shown in the figure). In a sense all varieties but A are absolute short-day plants at 20° and 25°C, but the picture is complicated because they also fail to flower on short day-lengths and might thus be considered day-intermediate. Most of them died on short day-lengths, however, but at least two remained alive but vegetative and could thus be considered true intermediates (especially variety F at 15°C). Work of this sort illustrates the points often brought out in the last chapter, that any present classification system can only be considered as highly tentative until many more data have been accumulated.

2. *Response Type and Plant Classification*

A legitimate question concerns the problem of photoperiodism and relationships within the plant kingdom. Are the response types each confined to certain taxonomic groups? Flowering response type symbols of the plants listed in the appendix were placed, along with the number representing the plant in question, on a chart of taxonomic relationships as envisaged by Lyman Benson, a taxonomist at Pomona College in California. It was very readily apparent that no simple relationship exists between plant classification and flowering response type. Virtually all of the response types, for example, may be found within one family, the Compositae. Furthermore, a given response type may be found in species occurring in many of the families of flowering plants, and some of these families may be on opposite ends of the classification scheme. In a few cases, a given response type seems to be rather predominant within a family, as for example the dual day-length requirements of many grasses, but in such cases it is not certain whether the relationship actually exists or whether we have simply not sampled the family widely enough to recognize the diversity of response types.

3. *Response Type and Evolution*

What does all this mean? Is short-day response in widely separated families, for example, a case of parallel evolution in which the same end point is arrived at from many different beginnings? The number of "convergences" make this seem rather unlikely. Two possible explanations for such an observation come to mind. First, it is quite possible that our categories do not represent single physiological mechanisms. Perhaps the processes which cause one species to be a short-day plant differ somewhat from those which make another species a short-day plant. In such a case the convergence would be apparent but not real in the physiological sense. There are examples which can be interpreted on this basis, as mentioned in the last chapter. Second, there are good reasons to believe that there is a fundamental mechanism common to virtually all higher plants, involving the pigment system (Chapter 7) and perhaps the flowering hormone itself (Chapter 9). This mechanism might be a highly plastic one, rather easily changed by mutations, crosses, etc. The response might become more specialized (e.g. day-length dependent) or it may become less so (day-neutral). Such modification by the forces of evolution might account for the many response types and differences within response type. Obviously the modification becomes quite extreme in some cases.

At present, we are not even sure whether day-length or temperature dependence might be considered primitive or advanced taxonomically. Surely with the inconsistencies mentioned above, it could never have any very significant meaning in this respect. It has been suggested that the day-neutral non-temperature sensitive condition is most primitive and might be found in some of the ancestral stocks, such as the buttercups, and that day-length and temperature sensitive species occur in the higher plant families, but in my attempts to correlate taxonomic relationship with flowering response type, I was unable to find any evidence for this viewpoint.

NATURAL SELECTION AND THE FLOWERING PROCESS

Natural selection is a powerful factor in populations of plants and animals. Thus any physiological or morphological feature of an organism might conceivably have survival value. On this basis we often ask ourselves, "What is the advantage of a given feature?" What, for example, is the advantage of being dependent upon some change in the environment for the initiation of flowers? How would such a response tend to insure survival?

It is rather easy to think of a distinct disadvantage. If a plant requires a certain day-length to initiate flowers, and only slight deviations from this day-length will either restrict the plant from flowering or cause it to flower at a season when it might be destroyed by frost, then the plant would seem to be restricted to a rather narrow distributional range. If it moves farther north or south, conditions may be unsuitable for growth and completion of the life cycle. Photoperiodism, then, might be a distinct disadvantage in that it restricts a species to a certain latitudinal range. How can it migrate and fill niches in the biotic community? What happens if a climate changes, but day-length does not? Throughout all the plant kingdom we encounter a great variety of highly ingenious dispersion mechanisms, apparently designed to allow a species to move freely on the surface of our planet. Yet a day-length requirement might defeat this end.

Certainly the picture is not so bleak. The plasticity of response type mentioned in the above paragraph must be one answer. For example, the plants studied by Mooney were all of the same taxonomic species, and yet they had become physiologically adapted to their given location. If we assume that they gradually migrated to their present location, they must have become physiologically adapted (by natural selection?) along the way. We can see in the Appendix, for example, that a given species may have many different response types, as for example with the chrysanthemum. Furthermore, a particular day-length requirement may not stop the plant from migrating, even though the requirement itself is not changed. Short-day plants which migrate southward will still encounter short-days, but at a slightly different time of year (somewhat earlier in late summer, for example). The new time of flowering may fall at a time of the year which is not particularly bad for the plant, and thus it will survive. Actually it is quite common to find a given species flowering at one time of year at one location and at a different time of year at another location.

The advantages of response to environment are also quite obvious. As mentioned above, a plant which requires an extended low temperature treatment for flowering will not burst into bloom during a warm spell in the late fall. Certainly a rather restrictive day-length requirement would allow a plant to fill a particular niche in time. Anyone who studies a given biotic community throughout a season will observe that there is a rather orderly sequence of flowering in the different species which make up the plant community. Certain species are early flowerers, others flower in mid-season, and some flower late in the year. Thus each one seems to fill its niche in seasonal time. If this does nothing else, it should at least serve to keep the honey bees on a rather sensible annual work schedule.

Perhaps the most distinct advantage of a rather critical day-length requirement is that such a requirement insures that all members of the population will be in bloom at about the same time. This is quite essential for efficient cross pollination among members of the population. Such intermingling of the germ plasm within a population seems to be distinctly advantageous in nature, although no attempt is made here to discuss the reasons for this.

It is easy to see how accurate day-length requirements might be developed within a natural population, since there is such a selection pressure against flowering out of turn. A plant which requires cross-pollination for fertilization is of course doomed if it flowers by itself. It will not transmit this anti-social attitude to any offspring! Thus the hard facts of survival dictate a rather close organization within natural populations, and there is little room for the non-conformist, who survives only with difficulty and may depend for his survival upon at least one other like himself. Yet, as mentioned above, it might be the non-conformist who provides the plasticity for migration into newly available niches, or into more suitable situations at times of climatic change. Progress could depend upon him. Of course, we will refrain from drawing any sort of human moral from all of this!

CHAPTER 4

THE LOW TEMPERATURE PROMOTION
OF FLOWERING

It is now known that flowering is induced or promoted by low temperature in a great many species of plants (see Chapter 2 and the appendix). As might be expected, there are many manifestations of this response, and physiological and descriptive investigations relating to the phenomena extend into the hundreds. Thus a brief summary limited to a single chapter in a book which emphasized the response to light must be only a review of some of the high points. A number of excellent review articles have recently been written, however, and the interested reader is urged to consult these summaries, where original papers are cited in profusion. In the preparation of this chapter, I have utilized the reviews of Napp-Zinn (24), Purvis (31), and Hartsema (17) in Volume XVI of the *Encyclopedia of Plant Physiology* (5), the review by Chouard (13) in Volume XI of the *Annual Reviews of Plant Physiology*, and a less available mimeographed copy of a talk given by Melchers. Work with gibberellins has recently been reviewed by Lang and Reinhard (56).

Researches relating to the responses of plants to cold caused a gradual accumulation of so much descriptive information that the terminology used to refer to the responses has become complicated and difficult, as each author may suggest new terms and definitions relating to his own findings. In the other direction, a single term has been applied to a number of phenomena which are almost certainly different in a physiological sense. Thus the term vernalization has been used to imply virtually any "positive" plant response to low temperature or any early flowering in response to any environmental variable! A majority of workers now feel that the term should at least be limited to low temperature promotion of flowering, and that it should not be used for other responses such as breaking of dormancy of certain seeds or buds. In addition, many of the scientists presently active in the field prefer to restrict use of the term to inductive promotion of flowering by low temperature. In Chapter 2 this practice was followed. First a distinction was made between causative effects of temperature on flowering (the response seems to be caused by the exposure to low, high, or alternating temperatures) and interactions of temperature with day-length (the photoperiodism response is modified to a greater or lesser extent by temperature). Second, a distinction was made between inductive (delayed) or non-inductive (direct) temperature effects on flowering. The term vernalization was restricted to causative, inductive promotions of flowering by low temperature. The term was first formulated in Russia as "jarovization" (jarovizacija). It is based on Latin, and an English equivalent might be "summerization". Vernalization as used in German and English implies "springization", as does printanization in French. Of course the implication is *not* that the genetics of the variety is changed to the summer or spring form, but only that the variety is made to *act* like the summer or spring form in response to an artificial cold treatment.

In this chapter we will review some of the history of the topic, summarize the principal facts relating to vernalization, discuss some of the recent work with applied pure chemicals and plant extracts, consider some theoretical aspects of vernalization, and examine briefly some work on direct (non-inductive) effects of temperature on flowering.

SOME HISTORICAL BACKGROUND PERSPECTIVE

The history of research in this field is worthy of summary, because it provides an excellent example of how quite obvious phenomena may be overlooked or simply considered uninteresting, either because research is being carried on more actively in some other field or because interpretation is held back by the predominance of some incorrect theory which is held in high regard. Thus as our understandings of chemistry were long retarded by the phlogiston theory, so knowledge about the environmental stimuli which might lead to flowering was held back by preoccupation with the idea that flowering came about in response to changes in plant nutrition.

Given spring wheat, which flowers a few weeks after planting in the spring, and given on the other hand winter wheat, which flowers by mid-summer only if it has been planted the previous fall, it would seem quite logical, perhaps even obvious, that the winter wheat must have been promoted in its flowering by the low temperatures of winter. Indeed, it did seem apparent to a number of people at least as early as the 1830's, and mention of the phenomenon can be found even before that. In 1849 the *New American Farm Book* described methods of treating moist winter wheat seed with cold so that plants will head rapidly when they are planted in the spring, and an anonymous report in 1837 also describes the process. J. H. Klippart described such methods in a report to the Ohio State Board of Agriculture in 1858. Other findings with various rosette-forming biennials as well as cereals appeared sporadically in 1875, 1881, 1898, 1899, 1903, 1906, and 1909, for example.

Some of these reports were made public by Georg Klebs, who spent the years before his rather early death at 61 (1918) at the University of Heidelberg. Klebs probably did more relating to the effects of environment upon plant growth and development than anyone else before Garner and Allard. He had a clear insight into the "specific structure" of the cell, which we would now call the genetic material and identify with deoxyribonucleic acid (DNA), and which he knew was in control of plant growth. He knew that the environment must act upon the internal conditions of the cell and thus influence the manifestation of the "specific structure" without changing the "specific structure" itself.

Klebs had studied flowering extensively. He had observed that sugar beets and many other biennials would remain vegetative for years when they were kept in a warm greenhouse (unless they were exposed for a time to near freezing temperatures), and also that some plants kept under incandescent lights during the night would flower while controls in the dark remained vegetative. Thus he had the basic information to formulate our present concepts of vernalization and photoperiodism. He was, however, a strong proponent of the nutrition theory, and when the vernalization response was clearly stated in 1910 and again in 1918 by G. Gassner, Klebs discussed the results on the basis of nutrition (see section on theory below), and his results with light were also explained in a nutritional context by considering the extra time for photosynthesis.

Thus it was Gassner who first introduced the concept of a cold requirement for flowering and attempted an analysis of the phenomenon, although his explanations were somewhat incomplete because he was unaware of the additional requirement for long-days in his species. Following the First World War there was considerable activity in Russia, and names such as Murinov, Maximov, Pojarkova, and Lysenko (or Lyssenko) appear in the literature. T. D. Lysenko's name is commonly associated closely with the early work on vernalization. His first paper appeared in 1928. Compared to the discoveries of many other workers, his contribution was probably rather slight (his finding that vernalized seed could be dried out and stored without losing the vernalized condition is most frequently mentioned), but he contributed the term Jarovization (1929) and he developed a theory, that of Phasic or Stadial Development, which has had a profound influence on Russian botany ever since. His departure from Klebs' clear insight, in that he thought the genetic material itself was influenced by the environment, is interesting in that it led to widespread political involvement in scientific research in the USSR (see section on theory below).

In Utrecht, Holland, A. H. Blaauw directed an active research laboratory concerned with flowering response of 23 "bulbous" and nine other horticultural plants, primarily as their flowering is influenced by temperature. Blaauw began the work in 1918 continuing until his death in 1942, and his laboratory is still very active. Blaauw's work is especially interesting because he used carefully controlled conditions long before our environmental chambers and phytotrons (see Chapter 5) were known. He was able to do this, because bulbs do not require light during the temperature treatment, and accurate control of temperature was even then technically quite feasible.

Beginning with the 1930's, an historical summary becomes difficult, because work was being carried out at many laboratories around the world. Nevertheless, the contributions of two laboratories are especially outstanding: G. Melchers and Anton Lang in Berlin-Dahlem and later Tübingen, Germany, and F. G. Gregory and O. N. Purvis in Imperial College, London. Any summary of current vernalization theory must rely heavily upon the work emanating from these laboratories. More recently, P. Chouard at the Sorbonne in Paris has studied many different plants and their responses to cold, as have workers such as S. J. Wellensiek, K. Verkerk, and others in Holland. W. W. Schwabe in England has pioneered work on short-day *Chrysanthemums*, and Klaus Napp-Zinn in Germany has developed a rather complex theory with *Arabidopsis thalinna*. Of course hundreds of other names appear in the literature of vernalization, but those mentioned above seem to recur most frequently or significantly.

Vernalization (12, 23, 24, 31)

Because of space limitation and the present complexity of the topic, the facts about vernalization will be summarized under eight headings. Of course, this is a rather drastic over-simplification, but the classical work of Melchers and of Gregory and their associates can now be outlined in a rather straightforward manner. Some of the newer work with other plants is also mentioned, although it does not always fall easily into this pattern. This is probably to be expected, and it does not invalidate the facts which were already available.

1. *The Many Response Types*

Melchers and Lang and other German workers studied two races of the henbane, *Hyoscyamus niger*. One race is an annual, flowering the summer following the spring in which it germinates. The other race is a biennial, which germinates one spring, grows as a rosette during summer, winters over, and then flowers the following summer. It had been shown as early as 1904 by C. Correns, the early German geneticist, that these two races differ from each other by only one gene. Both are long-day plants, but the annual has no cold requirement, while the biennial will remain vegetative indefinitely, literally for years, unless it has been exposed to a few weeks of very low temperatures. In the biennial race the cold requirement is absolute, and the long-day requirement is absolute for both races.

Purvis and Gregory studied primarily a spring and a winter race of Petkus rye, *Secale cereale*. These are both considered annuals, but one is a winter annual and the other is a spring annual. Both require long-days, and both will flower without cold, but the spring variety flowers in $7\frac{1}{2}$ weeks, while the winter variety requires 15 weeks. Cold treatment reduces flowering time in the winter variety to $7\frac{1}{2}$ weeks. The promotion by low temperatures can be replaced by a short-day induction, but this is not the case with henbane.

Of course, many other species have been used in studies on the mechanisms of vernalization. A number of these are listed in the appendix. They include such unexpected types as small summer annuals which germinate and flower within a few weeks and the common pea which flowers somewhat sooner (at a lower node) if germinating seeds are subjected to low temperature. A number of vegetables may be promoted in their flowering by low temperature treatment of the whole plants, but this could be a direct effect instead of vernalization in the narrow sense.

Most of the species used in vernalization studies also have a requirement for long-days following the low temperature treatment. The appendix shows that there are also a number of plants which require low temperature but are completely day-neutral following the low temperature treatment, and that certain varieties of chrysanthemum require low temperature followed by short-days in order to flower best. Nevertheless, no exception to the following generalization has yet been found: *plants which may be vernalized as seedlings have a subsequent long-day requirement.*

The genetics of cold requirement have been studied in a preliminary way. The situation is sometimes more complicated than the single gene difference discovered by Correns in *Hyoscyamus*, yet usually it is relatively simple. One of the more complex examples is wheat, in which the vernalization requirement exhibits a quantitative inheritance. The degree of cold requirement varies, and there are a number of steps or intergradations between an absolute requirement and no requirement at all. The vernalization requirement is commonly a double recessive; that is both genes for vernalization must be present if the trait is to be expressed. Perhaps this indicates that vernalization requirement is a defect in a gene which normally leads to flowering in the absence of cold. Thus when two defective genes are present, some sort of metabolism which takes place best at low temperatures is required for flowering. Of course two positive acting (instead of defective) genes could also account for the cold requirement.

At any rate the vernalization response is an operation of certain genes under certain environments. Thus it is not the flowering which is inherited, but it is the gene, the *potential to flower* under the right conditions.

2. *The Site of Perception of the Cold in Vernalization*

An early question asked which organ of the plant was involved in vernalization. The trick was to cool only a certain portion of the plant while leaving the rest at a normal temperature. It was discovered with plants beyond the seedling stage that cooling the leaves or the roots had no effect unless the apical meristem of the shoot was cooled. Such plants include onion, beet, henbane, and chrysanthemum. With a cereal seedling, the embryo itself perceives the cold, and not the endosperm. Thus it has been generalized that it is always the meristem that responds to cold in vernalization.

It was learned in 1955, however, that cotyledons of *Streptocarpus* could be vernalized, and in 1961 Wellensiek found that detached leaves of *Lunaria biennis* could be vernalized without buds (37). Such leaves regenerate flowering plants, but controls not subjected to cold produce only vegetative plants. In subsequent work he was able to show that the leaves had regenerating tissue at their base, and thus it was possible to suggest that vernalization requires dividing cells if it is to be effective (see Chapter 10). Response of seeds below 0°C suggests, on the other hand, that dividing cells are not essential. To settle the problem, careful cytological work will have to be done.

The meristematic cells of the growing point are changed by the cold in some way which allows them to become reproductive at a later date. The time between perception of the cold and actual expression of the stimulus by appearance of flowers might involve a fairly large portion of the life cycle. This is quite apparent for the cereals, which respond to cold in the seed stage. Even in henbane, if the cold treatment is followed by short-days the plant will remain vegetative for an indefinitely long period of time (190 days in one experiment for example) until it is finally converted to the reproductive condition by treatment with long-days. The meristem has progressed towards flowering in response to the cold, but it produces only leaves until long-days (presumably a stimulus sent from the leaf in response to long-days) complete its transformation to the reproductive state. True vernalization, then, provides one of the best known examples of the induced state, a topic to be discussed again in Chapter 10.

3. *Condition of the Plant*

As indicated in Chapter 2, the age of the plant at sensitivity depends upon the species. In cereals, the damp seed itself will respond to the cold, and immature seeds still on the mother plant have been vernalized. Cereals usually are vernalized in the seed or seedling condition, but it has been shown that mature cereal plants will also respond to the cold. This involves the technical difficulty of demonstrating a promotion by cold even though the plant would flower anyway within a short time. With this difficulty in mind, experimental results can be interpreted to show that vernalization is effective (to some extent at least) right up to the time the plant would flower anyway. The biennials as a rule do not respond in the seedling or moist seed stage. A rosette of leaves must be formed. *Hyoscyamus* must be at least 10 to 30 days old before it will respond to the low temperature treatment. Sugar beets normally respond when they are $2\frac{1}{2}$ to 3 months old, but the time can be shortened considerably if plants are exposed to continuous light during the low temperature treatments. In some plants (beet, henbane) sensitivity to vernalization increases at least for a time, as plants get older. In plants that respond as seedlings, however, the sensitivity may decrease as the plant ages. The topic of "ripeness to flower" is discussed again in Chapter 6.

In order for the cold to be perceived, the cells at the apical meristem can not be metabolically inactive. Completely dry seed will not respond, although seeds which have imbibed water only slightly are sensitive, even though they have absorbed insufficient water for active germination (30% to 40% produces ample sensitivity in wheat seeds). It would be interesting to know if this requirement for some water is related to an initiation of cell division, but so far such information is not available. Following vernalization of this slightly moist seed, Lysenko showed that it may be dried out again, stored, and mechanically sown. Thus the results of vernalization are relatively stable, once the process has taken place.

4. *Effective Temperatures and Times*

In rather early experiments, vernalization was carried to a maximum, and it was found that the temperature optimum was very broad. As shown in Fig. 4–1, temperatures from 1 to 7°C, are almost equally effective, and temperatures as high as 9 to 15°C have some effect. Temperatures a few degrees below freezing are also effective. In later experiments, a shorter vernalization time was used, and the temperature optimum was found to be slightly more sharp. That is, the curves had a peak (although not a very sharp one) at about 6°C. As we might expect, the exact optimum temperature depends upon the species.

As also might be expected, the optimum times for cold treatment

F

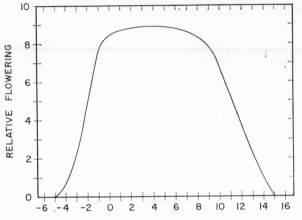

FIGURE 4-1
Final relative flowering response as a function of temperature during vernalization. The data represents response of Petkus rye to a 6-week period of treatment. Original data of Purvis and of Hänsel (see 31).

also vary for different species. In a few cases such as summer wheat which has only a quantitative vernalization requirement, 3 weeks seems to be an optimal time. In many other cases (e.g. henbane) the optimal duration may be longer than 3 months. In a few cases such as Petkus rye, an effect can be detected after only 4 days, but with some plants 4 to 8 weeks are required for any effect at all. Figure 4-2 shows results with Petkus rye. Usually treatment durations longer than the "optimum" produce the same effect as the optimum, but in a few instances there seemed to be a slight reduction in flowering with extremely long treatment times, and this was called oververnalization.

5. Devernalization

If the cold treatment is followed immediately by high temperatures, plants revert to the unvernalized condition, in which the cereals flower only after a long time, and the biennials remain completely vegetative and will not flower even on long-days, unless they are revernalized. Fairly high temperatures such as 35°C are most effective, but in rye temperatures above 15°C begin to have a devernalizing effect (see Fig. 4-1), and in henbane the neutral

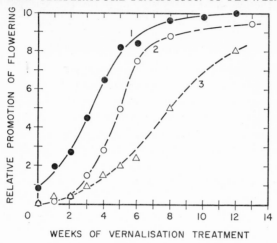

FIGURE 4–2

Final relative flowering response to different vernalization times. Curve 1 represents results with intact grain; curve 2 with excised embryos provided with 2% sugar; and curve 3 excised embryos without added carbohydrate. Data from Purvis (31). Compare the shape of these curves with the autocatalysis curves shown in Fig. 10–3.

temperature is about 20°C. Temperatures below this promote flowering, and temperatures above tend to nullify the effects of a low temperature treatment. The devernalizing high temperatures must be applied within a relatively short time after the cold treatment. Usually a period of 3 to 5 days is sufficient to stabilize the vernalized condition so that devernalization fails.

In some early experiments of the English workers, devernalization was brought about by drying and storing vernalized seed. Apparently this was because the seeds were not dried as thoroughly as they should have been. Nevertheless, the reversal to the unvernalized state was quite evident and beyond question.

6. *Vernalin*

In the laboratory of Melchers, it was possible to graft a vernalized henbane plant to a nonvernalized plant, causing the nonvernalized plant to flower. Thus Melchers suggested that a substance, which he called vernalin, was produced by the process of vernalization, and that this substance would pass across a graft union. Comparable results have now been obtained with a few species, but often this experiment fails.

In virtually all of these grafting experiments, the donor plant was flowering at the time the graft was made. In such cases it is possible that the product of the cold treatment was further elaborated into some final flowering stimulus such as the hormone produced in response to day-length (see Chapter 9), which will readily pass a graft union. Melchers performed one experiment, however, which cannot be interpreted in this way. He made a vegetative biennial henbane plant flower on long days, by grafting onto it a vegetative short-day tobacco plant (Maryland mammoth). The implication is that vernalin is a necessary precursor (in a physiological sense, at least) of the flowering hormone, and that flowering in henbane is blocked because this material is not present until after a cold treatment, while in tobacco the vernalin is present, but short-days are required to elaborate it further into the flowering hormone. It is a neat explanation, but we require more biochemical knowledge of what is going on before it can be accepted without reservation.

7. Substrate Requirements

A number of experimenters have demonstrated the need for oxygen in the vernalization process. Devernalization, on the other hand, can be carried out in a pure nitrogen atmosphere. Plants are converted back to the unvernalized condition, but they may not be damaged in any other apparent way.

The English workers have carried out a number of experiments with embryos from rye seeds, separated from the rest of the seed. These embryos may be cultured on an agar medium and their nutrient requirements can thus be investigated. Such excised embryos will respond to the vernalization treatment on a relatively inert medium (containing the essential mineral nutrients but no sugar), although the rate and extent of response is reduced. However, if the medium is supplied with a sugar such as sucrose, the response is delayed about 2 weeks but is otherwise similar to the response of intact seeds and seedlings (see Fig. 4–2). A number of sugars have been used and many of them are effective (for some unknown reason, certain ones are not). The details become involved, but it does seem clear that sugar or other carbohydrates, present in the storage parts of the seed, take part in the chemistry of the vernalization process (see Chapter 6 for related phenomena in photoperiodism).

8. *Interactions of Day-Lengths and Cold*

We saw in Chapter 2 that temperature and day-length may interact in all sorts of ways. Short-days will essentially replace the effects of cold in rye. Furthermore the response in rye is not absolute, since neither cold nor short-days are required for ultimate flowering. In *Campanula medium*, the response is absolute, and short-days and low temperatures are completely interchangeable. Either one or the other is required for flowering. A certain variety of radish is a long-day plant unless it is treated with cold, after which it becomes day-neutral. In this case, long-days and low temperature might be considered as interchangeable. A long-day variety of spinach has a critical day of about 14 hr unless it has been vernalized, in which case the critical day decreases to about 8 hr. Such temperature interactions could play an important role in future research on vernalization.

RESULTS WITH APPLIED PURE CHEMICALS AND PLANT EXTRACTS

As mentioned above, there is some evidence for a positive flower-promoting substance which is formed in response to cold treatment. In Chapter 1 we also introduced the concept of a general flowering hormone (to be discussed further in Chapters 9 and 10). Obviously an important goal of plant physiologists is to discover the chemical nature of such growth regulating compounds and to understand the biochemistry of their formation and action. The auxins (stem growth hormones) were the first such compounds to be isolated and at least partially characterized, but in recent years other compounds have begun to yield to this approach. Thus we now know about cell division factors such as kinetin and the coconut milk factors with all their relatives, growth inhibitors such as the unsaturated lactones and naringenin from dormant peach buds, and an extracted potato tuber-inducing principle. Whether or not the flowering hormone itself has been extracted remains in doubt at the time of this writing (see Chapter 9), but three classes of compounds extracted from plants are known to replace the requirement for vernalization. Of these three, the gibberellins are by far the most important, and it is possible that more detailed work could expand the gibberellin class to include the other two.

1. *The Soak Water from Vernalizing Seeds*

Harry Highkin (51) at the California Institute of Technology at Pasadena was able to speed up the flowering of certain pea varieties by treating the germinating seeds with low temperature. He then applied the water, which contained materials that had diffused out of the vernalizing seeds, to unvernalized pea seeds, or even to Petkus rye seeds, causing these plants to flower sooner. Gregory had obtained similar results years before with the soak water from vernalizing rye seeds, but his work was not continued. This work is of interest for two reasons: first, the extractions were made in direct connection with the vernalizing process, and second, preliminary analyses of Highkin's extract seemed to indicate that the active principle was a nucleotide. Furthermore, ribonucleic acid (RNA) has been reported to promote flowering of cold requiring plants. In a morphogenetic process such as flowering, the nucleic acids must participate, if only in the cell division phase of the process, and any evidence which implies that they might be active in other steps of the sequence of events must be viewed with interest.

2. *The Gibberellins* (56)

These compounds, originally extracted from a fungus, have been investigated very extensively during the past decade for their effects upon higher plants. The most common observation is that they are powerful promoters of stem elongation. Although there is usually no increase in dry weight of the plant, the stems may elongate up to six times as fast after gibberellin treatment, and in some cases such as rosette plants, treatment may cause elongation where none would have occurred otherwise.

It was discovered at an early stage of the research that some plants which normally require cold treatment or long days for flowering will flower under non-inductive conditions when they are treated with gibberellins (see Fig. 4–3). The response is widespread but of course varies considerably from species to species. As a rule repeated, substantial doses are required. For example, with *Samolus parviflorus* a total of 20 micrograms (1 μg/day for 20 days) was very effective, and 100 μg gave maximum effect. On the other hand, about 900 μg (5 μg/day for 6 months) was only partially effective on parsley. It is important to note that only 0.001 μg/plant will give a significant stem elongation response, and thus the doses required for promotion of flowering are exceptionally high.

The most impressive thing about this work is that a well-defined pattern is evident in the results with many species:

A. Flowering of many cold requiring plants with the rosette form is promoted in the absence of cold treatment by application of gibberellins, showing that gibberellins will substitute for a cold requirement. If the plant also has a long-day requirement, effects of gibberellins are best observed when plants are treated with the chemical under long-day conditions. In a few instances such plants have been made to flower with gibberellins in warm, short-day conditions, but often this results only in stem elongation but no flowering.

B. Long-day plants without a cold requirement may often be made to flower under short days by treatment with gibberellins. In a long-short-day species of *Bryophyllum*, application of gibberellin under short days promptly results in flowering, but application under long days does not. Thus gibberellins can substitute for the long-day requirement.

C. Applied gibberellins do not cause flowering of short-day plants under long-day conditions. There are a number of reports of promotion in flower development by gibberellins in short-day plants when the plants have been induced to flower by a short-day treatment, but gibberellins by themselves do not seem capable of substituting for short days. In some cases gibberellins inhibit the flowering of short-day plants, but the particular response seems to be strongly dependent upon the environment.

D. There are a few cases where applied gibberellins do not cause flowering of cold- or long-day-requiring plants. This is especially true of such plants with a caulescent growth form (an elongate stem instead of only a rosette of leaves). Indeed, virtually all short-day plants are of this type. Gibberellin treatment may also inhibit flowering of Petkus rye and other cereals when it is applied to young plants, although application to older plants may promote flowering. Rosette plants in which lateral buds become reproductive while the terminal bud remains vegetative (e.g. *Geum urbanum*) also seem resistant to gibberellins so far as flowering is concerned, although in some instances stem elongation is promoted. It seems clear, then, that gibberellins will often substitute for a vernalization or a long-day requirement (but seldom for both in the same plant) but not for a short-day requirement. It is possible that some of the exceptions might be due to experimental factors. Perhaps the dose was not high enough or applied for a long enough time. Or perhaps the proper gibberellin was not used. Recently Marian Michniewicz and Anton Lang (63) at Pasadena, California, have tried nine different gibberellins on five species of cold-requiring and long-day plants. They found a marked difference in response to these compounds. GA-7, which is relatively new and difficult to obtain, was highly effective in all species, including one which was known to be resistant to the more common gibberellic acid (GA-3). Furthermore, compounds which were effective on one species were in some instances completely ineffective on another species. GA-8 was ineffective in all five species. It is quite obvious, then, that work with a single form of gibberellin should not provide a basis for generalizations.

A primary question concerns whether the gibberellin effect is physiological or pharmacological. To paraphrase Lang and Reinhard, is gibberellin a picklock that will open a door for which it wasn't actually designed, or is it the natural key to flowering of cold-requiring and long-day plants?

There are at least two lines of evidence which indicate that gibberellins do indeed play a role in the flowering process under natural conditions. First, they are now known to occur in higher plants, perhaps all higher plants, as well as in the fungi. Although five of the gibberellins used by Michniewicz and Lang were isolated from the fungus *Fusarium moniliforme* (GA-2, GA-3, GA-4, GA-7, and GA-9), three were isolated from immature bean seeds (GA-5, GA-6 and GA-8), and one was found in either source (GA-1). Gibberellins extracted from a number of higher plants have now been used successfully to induce flowering. Perhaps most impressive of all, extracts from plants which are flowering are more effective than extracts from the same vegetative, cold-requiring or long-day species.

This brings us logically to the second evidence, which is that the gibberellin status of a plant changes markedly during induction by cold or long days. Lang has shown that both the quantity and the kinds of substances showing gibberellin activity change in various species during the induction period.

For many years after the hormone concept of flowering was developed, it was felt that proof of the theory would be the demonstration that flowering plants could be extracted, and the extracts applied back to vegetative plants causing them to flower. Extraction of vegetative plants should yield inactive (or weaker) extracts. This has now been clearly achieved with the gibberellins. Are the gibberellins, then, identical with the flowering hormone? Obviously not for short-day plants. It was suggested that there might be two hormones, one lacking in vegetative cold-requiring or long-day plants (gibberellins) and one lacking in vegetative short-day plants. Since a flowering short-day plant will cause a vegetative long-day graft partner to flower, both hormones would have to be required for flowering; but if that is the case then a vegetative short-day plant on long days, having the gibberellin but lacking the short-day product, should cause a vegetative long-day graft partner to flower on short days. This seldom occurs, but the above grafting experiment of Melchers is nearly an example.

Could the vernalin of Melchers be gibberellin? This does seem possible, although there are also a number of difficulties here. Treatment with gibberellin leads first to active cell division in the subapical region of the rosette axis, then to the formation of an elongate axis, and only secondarily to flower formation. Flowering in response to cold or to long days, however, occurs before or concurrent with stem elongation. Perhaps the activities associated with elongation of the stem lead to flower formation, and gibberellins only cause the stem elongation. Even this theory is not completely satisfying, but it does seem to be the most likely explanation presently available. Production of gibberellins in response to cold- or long-day could at least be *part* of the flowering process; a part capable of making the entire process begin if it becomes predominant.

Very recently, Jan A. D. Zeevaart and Lang (81) at Pasadena have performed an elegant series of experiments which demonstrate that in *Bryophyllum*, the long-short-day plant mentioned above, application of gibberellin leads to the formation of flowering hormone on short days but is not the hormone itself. Plants which had been induced to flower by application of gibberellin on short days would induce vegetative graft partners to flower even on long days. Since these plants will not flower on long days alone, even when treated with gibberellins, the effect could not be due to residual gibberellins coming from the donor plant.[4] Furthermore, the receptor plant could be used successfully as a donor in a second graft, and leaves which had grown out after the induction treatment could also be successfully used as donors. Thus it seems clear that in this plant gibberellins are capable of substituting for long days, but that short days are also required for elaboration of the flowering hormone. Once made, the hormone will induce grafted receptor plants through at least two graft "generations" (see Chapters 9 and 10 for further discussion of grafting experiments).

Gibberellins seem to be closely involved with the biochemistry of flowering in cold-requiring and long-day plants, and we can await many interesting developments in this field. At the same time it should be clear that the entire story is rather complicated. Evidence from extraction and reapplication experiments turns out to be somewhat less straightforward than was originally imagined.

3. *Substance E.*

H. Harada and J. P. Nitsch (29) at the Phytotron at Gif-sur-Yvette, near Paris, have obtained some very interesting results with extracts of a number of cold-requiring and long-day plants. On their chromatograms a peak of growth regulator activity which they called substance E appeared beginning at the time of stem elongation in bolting plants. Using a whole field of hollyhocks and a year of laboratory work, enough of this substance was obtained for a number of physiological experiments. It caused bolting and flowering of long-day *Rudbeckia speciosa* and *Nicotiana sylvestria* plants, as well as a cold-requiring variety of Japanese chrysanthemum.

As in the case of the soak water extracts and gibberellins extracted from vernalized plants, substance E is of interest because of its close correlation with the vernalization process. It could well be that cold treatment, for example, causes formation of substance E which then causes bolting (is it thus identical with vernalin?).

Substance E is also interesting because chemical analysis shows that it does not have the molecular structure common to GA-1 to GA-9 (not enough material was obtained to completely identify it

---

[4] H. Haranda at Gif-sur-Yvette near Paris has obtained very similar results with cold-requiring varieties of *Chrysanthemum morifolium*.

chemically), although it has many of the biological properties of the gibberellins, such as activity in the dwarf corn bioassay. It is inactive in certain auxin tests (*Avena* curvature) but active in other typical auxin responses (growth of sunflower tissue cultures). It will be highly interesting to follow future work with this substance.

THE THEORETICAL APPROACH TO VERNALIZATION

For many years now attempts have been made to discuss the mechanisms of vernalization in physiological terms. Three basic approaches to the problem are briefly summarized as follows:

1. *Antagonism Between Vegetative and Reproductive Growth*

Before any sort of formal work had begun on either vernalization or photoperiodism, the theory had begun to develop that flowering and vegetative growth are antagonistic to each other. It was deduced that any means of repressing vegetative growth would result in flowering. The general idea is still a part of much horticultural practice, and often such a response may be observed. Plants are not watered, not fertilized, pruned heavily, girdled or otherwise mutilated to promote flowering. It was easy to apply the concept to vernalization by saying that the inhibition of vegetative growth by cold allowed reproductive growth to gain the upper hand.

Yet as is often the case with such generalizations, the concept proved to be highly over simplified. It only holds true for certain plants. It is equally true for at least as many plants, including many which require vernalization, that conditions which promote rapid vegetative growth also promote flowering. The argument raged for years around the concept of the carbohydrate/nitrogen ratio. It was said that low nitrogen (obtained by controlling fertilizer application) allowed a high development of carbohydrate and resulted in flowering. Low nitrogen does seem to promote flowering in many species (e.g. certain fruits), but this is not true for many other plants, where the opposite condition seems to hold. For example, cocklebur is strongly promoted in its flowering by high nitrogen levels in the soil — providing that the proper day-length conditions have first been met (Table 5–1). Thus there are no broad experimental grounds for the idea that repression of vegetative growth leads to flowering.

2. *The Hypothesis of Phasic Development*

In Chapter 6, we shall discuss "ripeness to flower". The concept states that a plant must attain a certain stage before it is capable of becoming reproductive. It was formulated by Klebs in 1918, and following work on vernalization, it was used as a point of departure by Lysenko for the theory of phasic or stadial development. In its simplest form, the theory states that plants must progress in order through a series of developmental stages, each subjected to environmental control. In addition, the theory now includes the restrictions that the stages are firmly set, the same in all species, absolutely irreversible, and due to physical changes of the protoplasm itself. Such rigid requirements should, of course, make the hypothesis quite simple to test, and most evidence seems negative. First, the many response types described in Chapter 2 make us sceptical of the theory's universal application. Second, devernalization, reversion to vegetative conditions under low intensity light, or after defoliation (see Chapter 10), and the perennial habit itself make us highly sceptical of the part of the theory that says that the stages are irreversible.

Thus in the western world, the evidence has been examined and the theory rejected. In the Iron Curtain countries, however, Lysenko and others went on to deduce from the theory that our present concepts of genetics were inadequate. They returned to the discredited idea of modification of the genetic material by environment. I have so far been unable to follow the reasoning, but apparently it arose from the vernalization of winter cereals — the "conversion" of these varieties to the spring form. At any rate, Lysenko has since had varying degrees of influence on Russian biology (particularly genetics), and the science has never quite recovered (see 53).

To summarize, we would readily admit that plants go through a series of steps in their life cycle, but we see no reason to restrict this within the limits of a formal theory as narrow as the one under discussion. Certainly the steps in a life cycle may be determined by vernalisation or photoperiodism, but we see no simple theory which relates this to the growth and development of all plants. We need many more data before such a theory can be formulated — and all indications are that it will not even then be simple.

3. *The Hypothesis of Flower Producing Substances*

Experience in biological research teaches that it is fairly safe to

FIGURE 4–3

The effects of cold or gibberellins on the flowering and growth of a cold requiring plant (Carrot, Early French Forcing). Left: control maintained above 17°C; right: 8 weeks of cold treatment; and center: 10μg of gibberellins daily. The flowering plants are 1 m tall. Photograph used by permission (see Anton Lang, 1957, *Proc. Nat. Acad. Sci.* 43, 709–717).

think of plant or animal function in terms of biochemistry. We now know that photosynthesis, respiration, and many other plant functions are the result of chemical processes going on within the cell. This even applies to processes such as growth, which seem to be under the control of hormones such as the gibberellins and auxins. Thus it seems reasonable to imagine that vernalization might lead to some chemical change within the plant which in turn leads to flowering. This is the approach to an understanding of vernalization taken by most plant physiologists.

A problem immediately arises. How can *low* temperatures *promote* the accumulation of a compound? Chemical reactions are usually speeded up by increasing the temperature and slowed down by decreasing the temperature. The clue to the solution of the problem may prove to be present in the phenomenon of devernalization, in which high temperatures may inhibit flowering and reverse the effects of low temperatures. It was suggested in the laboratories of Gregory and of Melchers that two reactions were going on within the plant. One is a synthesis of flower promoting substances; the other a destruction of these substances:

$$A \xrightarrow{\text{I}} B \xrightarrow{\text{III}} D$$
$$C \xleftarrow{\text{II}}$$

Reaction I, the synthesis, may proceed even at low temperatures. Reaction II, the destruction, might go much more slowly at low temperatures, but it might increase with increase in temperature much more rapidly than reaction I. Thus at low temperatures B would tend to accumulate, but at higher temperatures B would be destroyed (converted to C — or perhaps back to A) before it could accumulate, as illustrated in Fig. 4–4. At normal room temperatures B is further converted to D, another step towards flower formation. D is not destroyed by high temperatures, so after a few days devernalization will not occur. Various other reactions may also be postulated to account for subsequent day-length requirements when they occur.

The theory is a neat one, but in a sense it is little more than a graphic representation of the experimental observations. How does it relate to present work with gibberellins and other extracts? We

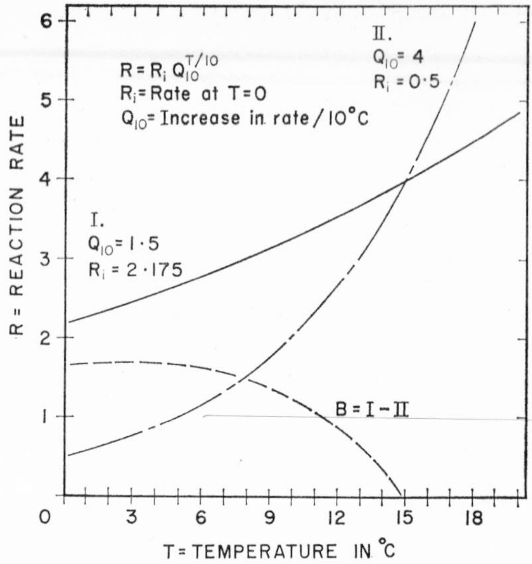
R = R_i Q_10^T/10
R_i = Rate at T = 0
Q_10 = Increase in rate / 10°C
II.
Q_10 = 4
R_i = 0·5
I.
Q_10 = 1·5
R_i = 2·175
B = I − II
R = REACTION RATE
T = TEMPERATURE IN °C
0
3
6
9
12
15
18
1
2
3
4
5
6

FIGURE 4–4

Sample curves showing hypothetical reaction rates as a function of temperature for reactions with Q_{10} values of 1.5 or 4.0. If the reaction with Q_{10}= 1.5 is considered to be reaction I in the formula in the text, and the reaction with Q_{10}= 4.0 to be reaction II, then the hypothetical product B will be proportional to curve II minus curve I, as shown. Compare the shape of curve B with Fig. 4–1.

might by definition identify B in the above formula with the vernalin of Melchers. Is the active fraction of the soak waters also vernalin? It seems possible, but too little work has been done to say. How about substance E? Here the evidence is surely in favor, but again more work is required.

What about the gibberellins? In this instance enough work has been done to allow the tentative conclusion that the gibberellins and vernalin are *not* identical. But if the gibberellins are nevertheless a physiological part of the flowering mechanism in some plants at least, as the evidence also seems to indicate, where do they fit in the above formula? Apparently they do not fit, and so our formula must be too simple. Certainly there is a great deal left to learn about the biochemistry of vernalization!

DIRECT EFFECTS OF TEMPERATURE ON FLOWERING

In a broad sense, vernalization could be defined as any promotion of flowering by low temperature, and such a definition would be as good as any if it were accepted by workers in the field. There is a current tendency, however, to limit use of the term to delayed or inductive effects (see above and Chapter 2), although it is certainly not easy to draw the line between direct and delayed effects in all cases. Following current opinion, the two cases of flowering response to temperature discussed in this last section are not to be considered as vernalization. As a matter of fact, flower formation in the bulbs often occurs in response to relatively high temperatures (and in one case at least, this is inductive!) and thus cannot be considered vernalization under any presently acceptable definition.

1. *Direct Flower Formation in Adult Plants*

There are a few instances in which mature plants must be exposed to low temperatures if flowering is to occur, and the formation of flower buds can be observed to occur during the cold treatment instead of later. Brussels sprouts (*Brassica oleracea gemmifera*) were studied in some detail by Pearl Stokes and K. Verkerk (72) in Wageningen, Holland, and according to these workers other species responding this way include cabbage, sweet turnip, foxglove (*Digitalis purpurea*), stocks (*Matthiola*), and Sweet William (*Dianthus barbatus*).

To begin with, response to the cold treatment is dependent upon the Brussels sprouts plants first reaching ripeness to flower. The presence of a juvenile phase is not unusual (see Chapter 6), but in this case Stokes and Verkerk found that they could recognize ripeness to flower by morphological changes in the bud. Following this stage, the plants are then ready to respond to 3 weeks or more of low temperature (3°C), after which flower primordia begin to develop. If no primordia are evident when plants are returned to higher temperatures, flowering will not. occur. The extent of flowering is dependent upon the length of the cold period (6 to 9 weeks produce full bloom), and there seems to be no interaction with day-length so far as flowering is concerned.

Of course, the extent of flowering will be modified by the rate of elongation (rate of bolting) of the stems after plants are returned to warm conditions. This response is also dependent upon the cold, and thus there is an aspect of flowering in this species in which an inductive effect can be observed. It seems reasonable to expect that this response is related to endogenous gibberellins.

2. *Flower Formation in Bulbous Plants* (17)

Blaauw and his team of workers in Holland were primarily interested in discovering the optimal storage temperatures and other treatments which might be used to produce an abundant harvest of commercial flower bulbs and then to insure the best growth, flowering, and sometimes accelerated blooming of these plants after they were turned over to the customer. Often very special treatments had to be used, such as in the shipment of bulbs to the Southern Hemisphere. Their approach was to carefully examine the internal morphology of the bulbs during an annual cycle in the field (sometimes at the native location) and during commercial production. Then they would subject bulbs to accurately controlled storage conditions, still observing changes in morphology, and finally bulbs were planted in the greenhouse or the field and their ultimate growth and flowering observed. Such a procedure has been going on for more than 40 years, and we now have available an extensive literature relating to the subject. Of course, work has also been done in other laboratories, especially during more recent years.

Unfortunately, the work has been almost completely descriptive, since this serves the ends of the supporting industry very well. Relatively few attempts have been made to study the biochemical changes taking place during flower formation or to formulate any sort of theoretical scheme attempting to understand the nature of the changes. Perhaps this is not so unfortunate after all, because we at least have an extensive foundation of descriptive data for future formulation of theory.

A few generalizations are apparent from the work. Flower formation in the bulbs is first dependent completely upon their size. Bulbs which are too small will not respond. In some cases (e.g. tulip) all leaf primordia are formed before flower formation, in other cases (e.g. iris) the number of leaves formed before the flowers is variable, and in other cases (e.g. hyacinth) leaf and flower formation may go on simultaneously. All species studied showed optimal temperatures for flower formation, and temperatures too high or too low either retarded or completely inhibited flowering. In most bulbs, the optima for flower initiation were relatively high, but sometimes fairly low temperatures were most effective (see below). Low temperatures might also be concerned with breaking of a "rest period", or they might be required for optimum stem elongation after flowers begin to emerge (another gibberellin effect?). The actual temperature responses were closely correlated with the behaviour of the plant in the field, and Hartsema divides these response types into seven categories. These have been condensed and simplified somewhat into the following four groups:

A. Flower primordia form a year in advance, before bulbs can be harvested. Obviously effects of bulb storage temperatures on flower initiation cannot be studied in these species, since flowers are formed long before bulbs can be stored. Photoperiod and temperature conditions could be important during initiation, however, but these factors have seldom been studied. In two tropical species, leaf and flower formation for the next year go on simultaneously during the entire assimilation period (*Hippeastrum, Zephyranthes*). In other species flower formation for the next year also occurs before this year's flowers are gone, but in a somewhat more regular fashion (*Amaryllis belladonna, Nerine sariensis*). In others formation occurs after blooming but before the bulbs can be harvested (*Narcissus* or daffodil, *Galanthus, Leucojum, Convallaria*).

B. Flower primordia form during the storage period after harvest in the summer but before replanting in the fall. Perhaps the most thorough work has been done with plants in this category. A typical curve showing optimal storage conditions for tulip is shown in Fig. 4–5. Examples include *Tulipa, Hyacinthus*, and *Crocus*. Optimal temperatures for flowering are relatively high (17–20°C for tulips, 25.5°C for hyacinths), although low temperatures are required for early blooming and good stem elongation.

C. Flowers are formed during the winter. If bulbs are planted in the fall (most bulbous irises — see Fig. 4–5), then flower formation occurs during the low temperatures of winter (9–13°C optimum), but it was found that a high temperature (23–30°C) pretreatment was essential if flower formation was to occur at all. This seems to be a true inductive or delayed effect, very similar to vernalization except that high temperatures are required instead of low. If bulbs are stored during the winter, then flower formation occurs during storage,

G

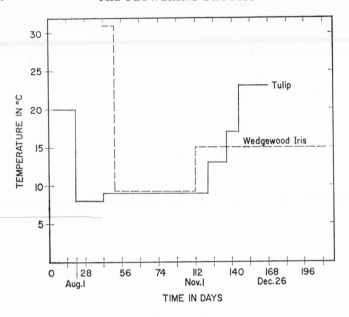

FIGURE 4–5

Temperature treatment for early flowering of *Tulipa gesneriana* " W. Copland " and of *Iris xiphium* " Imperator ". With the tulip, flower initiation begins and is well under way during the 20°C treatment. Moving to 8° and then 9° provides an acceleration in blooming, so that flowers are produced at Christmas. Continuous 9°C treatment gives equal earliness, but quality is poor unless the 20°C treatment is given first. The bulbs are planted in a controlled temperature greenhouse about midway during the low-temperature treatment. The temperature is first raised when the leaf tips are visible, then again when they are 3 cm long, and finally again when they are 6 cm long. The low-temperature optimum and the gradual change to higher optima as active growth begins have been related to conversions of starch to sugar during this period. Respiration rates also increase. With iris, the short period at high temperature is completely essential to flowering, although actual initiation of flower primordia does not occur until the bulbs have been moved from low temperature to 15°C, at which time the sprouts are about 6 cm long. Again, the 9°C treatment is to insure earliness. At temperatures much above 15°C during the last part of the treatment, abnormal flowers may be produced. Low light intensities will also result in " blasted " flowers at this time, especially if the temperatures are not right. If extremely high temperatures (38°C) are used during the first flower induction period, flower parts are increased or decreased, or tetramerous, pentamerous, or dimerous flowers result. Data from Annie M. Hartsema (17).

perhaps continuing after planting in the spring. Blooming follows rather shortly after flower formation. Examples are *Allium cepa*, *A. escalonicum*, *Lilium*, and *Galtonia*. In this intermediate group, optimum temperatures may be rather low (onion, 13°C) or somewhat higher (lily, 23°C).

D. Flower primordia form after bulbs are replanted in the spring. Examples include *Gladiolus*, *Freesia*, and *Anemone*. Storage temperatures seem to have little influence here, and temperatures during formation are difficult to study since initiation and flowering occur in rapid succession in the field after planting.

Obviously our most important present assignment so far as temperature effects on flowering are concerned, is to begin to gain some insight into the physiological and biochemical mechanisms which control these processes. There is still much descriptive background to be obtained, but we now have a sufficiently broad foundation that studies on mechanism should begin to be more fruitful than in the past. These bulbs which are small enough to handle easily but large enough to work with conveniently, and which are uninfluenced by light, might offer an excellent opportunity for research along these lines. It should, however, be quite apparent by now that any desire to understand the operation of a single plant will probably have to be satisfied by work on that particular plant, and findings with other subjects can at best only show the way.

CHAPTER 5

METHODS OF EXPERIMENTATION
WITH COCKLEBUR

HAVING completed the survey of some principles of photoperiodism as they apply to plants in general, we are now ready to begin our discussion of the mechanisms involved in the initiation of flowers, concentrating somewhat on a single species: the cocklebur. The usual procedure is to cause flowering of this short-day plant by exposing it to an uninterrupted dark period, carrying out various treatments in conjunction with this, and then studying the subsequent extent of flowering in treated and control plants.

There are many variations upon this theme; many treatments which may be given, and many details which must be considered. The primary purpose of this chapter is to provide insight into this experimental approach. I will describe the facilities and procedures which we use in our laboratory at Colorado State University, along with some of the methods used by other workers in attacking problems which we have not studied in Colorado. While our methods are special in certain ways, they are representative of the basic approach to photoperiodism, and as such they might serve as a framework for evaluation of the experiments described in the rest of the book.

In past years I have received a number of letters from high-school students asking how they might carry out experiments in photoperiodism. The following chapter is not written as a cook-book for such undertakings, but a little bit of the ingenuity displayed by such students should easily adapt the principles to amateur study.

I. GROWING THE PLANTS

1. *Facilities for Growing Plants*

Figure 5–1 is an interior view of our greenhouse at Colorado State University. Temperature control by evaporative cooling (as in the figure) or air conditioning might be expected to improve the experimental results. Flowering response of cockleburs is fairly sensitive to light intensity, and they respond best under full sunlight, so it has been our practice not to apply shading to the greenhouse glass during summer. The higher temperatures seem preferable to low light intensities.

Detailed studies on the effects of temperature and certain other factors require better environmental control than that found in most greenhouses. Thus controlled environment growth chambers or rooms are coming into more frequent use. A number of such installations constituting a single facility is referred to as a phytotron. In any case the expense is very high, the light intensities are relatively low (compared to sunlight), light quality is a serious problem, and yet the results obtained with such facilities seem to make them well worth the cost, as indicated by a number of experiments presented in this book. Figure 5–2 shows one of our eight chambers at Colorado State University.

In some cases the physical facilities for studies on photoperiodism deviate widely from those described here. The duckweed (*Lemna perpusilla*, 6746), for example, is a short-day plant which has been studied in some detail by W. Hillman (3) then at Yale University. All of his studies are carried out in the laboratory with the small plants floating on nutrient solutions.

2. *The Seeds*

Seeds of many species sensitive to day-length may be purchased from local dealers. In North America, it is usually possible to collect cocklebur seeds from weedy waste places in the fall of the year. The seeds used in most studies in the United States were originally collected near Chicago, Illinois, and they have since been propagated by workers at the University of California at Los Angeles, the California Institute of Technology in Pasadena, and other locations. Cockleburs collected in more southern latitudes (around Los Angeles for example) are sometimes not uniformly sensitive to photoperiod. The days are not as long in summer, and apparently plants have not been as carefully selected by nature. Various other plants sensitive to photoperiod are also weeds and may be collected in nature.

Seeds may be planted directly in the pots, perhaps two or three to the pot, thinning to a single plant after germination. We germinate the seeds in flats, subsequently transplanting them to pots, making more efficient use of the available seeds. It is probably best to plant cockleburs in sand, as the sand tends to pull the bur from the seedling during germination. In the early days of work with cocklebur it was thought that the seeds should be removed from the bur, or that the seed should be scarified by trimming the horns off the end of the bur. Such procedures may improve germination somewhat, but they are much more trouble than they are worth. We plant seeds directly with no previous treatment. Soaking the seeds for 24 hr before planting makes them germinate exactly one day sooner after planting!

3. *Control of Day-length*

With short-day plants the day-length must be artificially extended to maintain the plants in a vegetative condition prior to the experiment. We do this with incandescent lights in reflectors (see Fig. 5–1). Time switches turn the lights on about 4.00 p.m., in time for the gardener to check them and replace burned out bulbs, and off about 1.00 a.m. Plants remain vegetative on dark periods of 8 hr or less, but as a safety factor we seldom allow them to receive a dark period exceeding 5 or 6 hr.

With some long-day plants incandescent light is best for extending the day-length to induce flowering. This is apparently due to the mixture of red and far-red light produced by incandescent bulbs (see Section III, 2, below). Some studies indicate that incandescent light may also be best for maintaining short-day plants in a vegetative condition, but fluorescent light seems to be quite adequate. Some greenhouses use fluorescent tubes placed along the structural bars supporting the glass, where they cause virtually no more shadow in the daytime than would the structure itself. They are also cheaper to operate, although the initial cost of installation is somewhat higher.

In greenhouses where short-days must be provided regularly (e.g. to maintain long-day plants in a vegetative condition) black curtains are usually slid over a framework of pipe on the greenhouse bench. The curtains are often pulled over the plants about 4.00 p.m. and removed at 8.00 a.m.

4. *Temperature*

Temperature is relatively unimportant in photoperiodism studies with cocklebur. No temperature treatment yet discovered will cause flowering, but plants will flower on short days over a considerable

FIGURE 5–1

The cocklebur greenhouse at Colorado State University. Note cooling pads on the left, exhaust fan box on the right, lights hanging above the benches, their time switches on the wall to the rear, thermoregulator under a shield protecting it from the direct rays of the sun (on a post just right of center) plants trimmed and numbered for an experiment, and the cart on the concrete walk-way to the left. Note also plastic pots made by melting (with a hot soldering iron) two holes on opposite sides of pint plastic ice cream containers, and painting them with asphalt aluminum paint to reflect light and heat. Plastic is easy to handle, light, and doesn't absorb chemicals. Larger pots produce slightly better plants but require more soil and are heavier.

FIGURE 5–2

A growth chamber constructed at Colorado State University. Inside
dimensions are 4′ × 4′. Lights are 4′ Sylvania VHO fluorescent lamps
(26 lamps) and 100 Watt incandescent lamps (9 lamps). Temperature
control is obtained by modulating the rate of flow of chilled brine
(ethylene glycol solution) or hot water through cooling or heating coils
at the back of the chamber. The control system is pneumatic (Johnson
Service Company). As the temperature exceeds the setting on the
thermoregulator, pressure increases so that the normally closed chilled
brine valve begins to open, admitting cooling fluid to the coils. As
temperature decreases below the setting, pressure decreases and the normally open hot water valve begins to open. In case of pressure failure,
hot water is no longer supplied to the chambers; upon overheating,
lights are automatically turned off (except for a pilot lamp to protect
photoperiodism experiments), and a red light is turned on in the greenhouse headhouse where it can readily be noticed.

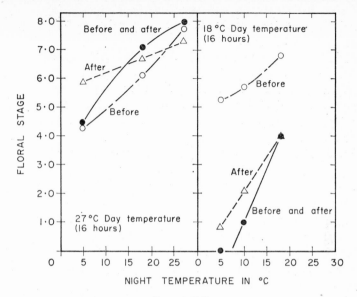

FIGURE 5–3

Effects upon flowering of cocklebur of various day and night temperatures applied before, after, or before and after a 16-hr inductive dark period at 23°C. The system of Floral Stages which indicates relative flowering response is explained near the end of this chapter (see Fig. 5–8 and Table 5–4). Experiments were carried out during the spring of 1962 in chambers such as that shown in Fig. 5–2.

range of temperatures. Yet in recent growth chamber studies, we have found that flowering is strongly modified in a quantitative way by temperature treatment. For years we have experienced difficulties, especially in winter, in exactly duplicating the results of two or more "identical" experiments. The results of our temperature studies, summarized in Fig. 5–3, give one possible explanation. Low night temperatures and low day temperatures are bad for flowering of cocklebur, whether they are given before the long dark period or following it. Even though the temperature is maintained fairly high in the greenhouse in the winter, plants may be radiating to the cold glass above, so that the temperature of the leaf is lower than air temperature alone might indicate. We have now obtained much better results by maintaining temperatures quite high, day and night, summer and winter, at high or low light intensities. This optimal esponse to high temperatures agrees well with the idea that our strain of plants is adapted to the hot summer days and nights which prevail around their native Chicago.

The best temperature conditions will obviously depend strongly on the species being used (Chapter 2). Probably an alteration in temperature between day and night is best for most plants, but we applied this principle to the cocklebur for a decade, and it was probably responsible for most of the difficulties encountered with our winter experiments! Of course, optimum temperature conditions should be determined for any new species to be investigated.

5.   Other Greenhouse Procedures

In our methodology we keep plants trimmed, so that only the relatively young leaves are left on the plant at any time. When plants have produced four or five leaves longer than about 1 cm, all but the first one or two are removed by trimming. This makes subsequent steps in our procedure easier, and it also seems to cause the plants to reach a mature stage at an earlier date.

Cocklebur plants flower in response to the long night much better if they are provided with an optimal supply of nutrients in the soil, as shown in Table 5–1. The uniformity of response is greatly improved,

TABLE 5–1.   EFFECTS OF FERTILIZER UPON FLOWERING AND UNIFORMITY OF FLORAL STAGES IN COCKLEBUR

Treatment[2]	Floral stages (relative flowering[1])	
	Plants fertilized[3] on Sept. 14, 1960, not again	Plants fertilized[3] on Sept. 14, 30, Oct. 3, 6, 12, 16, 20, 24, 1960
1. Distilled Water	2.7 ± 1.61[4]	5.1 ± 0.74[4]
2. 5.0×10^{-4} M Naphthalenacetic acid	0.9 ± 1.25	3.3 ± 0.48
3. 2.0×10^{-3} M Thiouracil	2.1 ± 1.45	4.3 ± 0.67
4. 8.0×10^{-3} M Thiouracil	0.7 ± 1.11	3.2 ± 1.23

[1] The Floral Stage system is described at the end of this chapter.
[2] Plants given one 16-hr inductive dark period, Oct. 31, 1960. Previously trimmed to young No. 3 leaf. Buds examined after 9 days. Plants dipped in solution containing 3 drops wetting agent (Vatsol)/250 ml just before the dark period.
[3] Fertilized with ⅛ teaspoon of Ortho-Gro 16–16–8.
[4] Standard deviations.

and the rate of development of the flowers is increased. The flowering response increases right up to the point where the plant begins to be damaged by too much fertilizer. Other studies indicate that phosphorus and nitrogen are especially important in improving the flowering response of cocklebur (25), but as mentioned in Chapter 4, some species flower best when they are limited in the amount of fertilizer that they receive.

We have found carts, such as that shown in Fig. 5–1, to be handy for most greenhouse procedures as well as for the experiments themselves. Our carts have four freely pivoting casters, which make them easy to move in any direction. Carts are commercially available, but we obtained the desired design only by having the carts built locally.

II. Preparing Plants for Experimentation

1. *Selection for Uniformity*

Plants must have reached an age of proper sensitivity. This seems to depend considerably upon growing conditions, but 60-day-old plants are almost always quite sensitive, and under optimum conditions, even 30-day-old plants will respond rather nicely. After plants are chosen for their general condition, age, height, etc., we sort them further according to leaf size. At an early stage of experimentation, I measured the length of the leaf midribs from the base to the tip of the leaf on 100 plants. The smallest leaf which was longer than 1 cm was called leaf No. 1, the next largest leaf, leaf No. 2, and so on. The average leaf lengths are shown in Fig. 5–4. Plants were then chosen on which all the No. 3 leaves were between 6.9 and 8.5 cm. In one representative group of these plants all of the leaves except No. 1 were trimmed off, in another all leaves except No. 2 were removed, and so on. Plants trimmed in this way were then given an inductive dark period, and their flowering was checked a few days later. The results are also shown in Fig. 5–4. The No. 3 leaf was clearly the most sensitive to photoperiodic induction.

It is often handy to use plants which have been trimmed so that only one leaf remains. This makes application of chemicals simpler, and it is especially helpful in light studies where intensity at the leaf level must be determined. We use plants trimmed in this way as a standard procedure. Since the No. 3 leaf is most sensitive, it is used

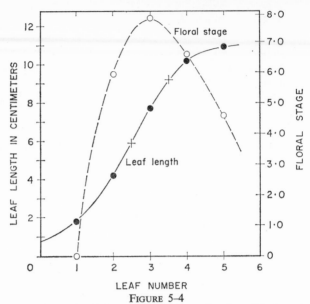

LEAF NUMBER

FIGURE 5–4

Length of the leaf midrib and flowering as a function of leaf number. The smallest leaf longer than 1 cm is considered to be number 1, the next largest number 2, etc. Experiments were performed in the fall of 1953 in Pasadena, California. See F. B. Salisbury, 1955, *Plant Physiol.* 30, 327–334.

almost exclusively. Plants are sorted with the help of a plastic measuring device, and the No. 2 leaf is removed along with all of those below the No. 3 leaf. We sort for No. 3 leaves which are 5.9 to 7.7 cm long (small), 7.7 to 9.2 cm long (large), or 6.9 to 8.5 cm long (typical). Since the No. 1 leaf is not sensitive, we usually leave it on the plant. Plants are normally trimmed one day before the experiment so that they can photosynthesize for one day without being shaded or disturbed. Sometimes plants seem to grow according to a curve different from that shown in Fig. 5–4. In such cases the No. 3 leaf as defined by our measurements may not be the most sensitive, but we use it anyway merely as standard procedure. If the No. 1 leaf appears large enough to be sensitive, we may remove it.

2. *Arrangement of Plants on the Bench*

Plants are first arranged on the bench so that all those of one kind are together. For example, all of the tallest or oldest plants may be placed at one end of the bench so that they do not shade the shorter ones. Plants are arranged after trimming so that the leaves all face in one direction and therefore do not shade each other. There is admittedly an element of superstition in this, but perhaps this is something to be recognized and used more often in science. We have not yet carried out detailed experiments to find out how important some of these procedures might be, but in the meantime we follow the procedures anyway.

In setting up the experiment, plants are numbered consecutively beginning on one end of the bench, in the manner illustrated by the diagram in Fig. 5–5. All of the plants with the same number receive

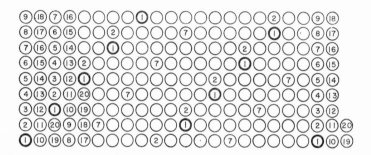

TWENTY TREATMENTS, TEN PLANTS PER TREATMENT

FIGURE 5–5

Method of numbering plants on the bench to insure that each group of plants receiving the same treatment is exposed to the full range of environmental variability. The arrangement illustrates an experiment with 20 treatments and 10 plants for each treatment.

the same experimental treatment. If in this numbering system, all of the No. 1 plants end up on the front of the bench, then we change the number of plants in the row or make some other adjustment so that the number ones form a diagonal across the bench as shown in Fig. 5–5. The idea is to try to expose each experimental treatment to all of the environmental variability which might be found on the greenhouse bench. Plants in the middle of the bench always grow taller, and one end of the bench may be closer to the heater, more shaded, etc.

This method of numbering provides a sort of "non-random randomness". Since the treatments do occur along diagonal lines, they are not truly random, and in a technical sense statistical evaluations cannot be applied to the results. We have found such an arrangement essential, however, since it is often necessary to find plants of a given treatment in a short time interval, and one cannot search through a truly random distribution. Furthermore, it would be quite amazing if the environment in the greenhouse varied in such a way that one treatment, with plants lined up diagonally, would have a significantly different environment than others in diagonal lines next to it. The distribution of plants may not be random, but it is quite different from the distribution of environmental variables on the greenhouse bench.

Statistics have seldom been used in the evaluation of experiments such as those described in the ensuing chapters, although use of simple tools such as the standard error of the mean has often helped to define the variability. As a rule, there is a clear cut difference between controls and treated plants — one may flower at a high rate and the other may not flower at all. As can be seen from many of the figures in this book, the experimental points (the actual data) do not always fall on the curves as drawn, but it is equally true that the trends are quite clear.

III. EXPERIMENTAL METHODS

1. *The Darkroom*

As mentioned earlier, dark curtains may be used on the greenhouse bench. This is a standard procedure for maintaining long-day plants in the vegetative condition, and of course short-day plants may be induced in this way. With the cocklebur, however, a single long dark period is sufficient, and in most cases, it is quite advisable to only use a single dark period,[5] in which case plants may be wheeled into a dark room where temperature and humidity can be controlled. In our experiments we usually maintain the temperature around 23°C, which is near the optimum for flowering, although temperatures from 20 to 30°C are quite satisfactory. If chemicals in solution are applied to plants, it is advisable to maintain a high humidity, since this seems to help in penetration of the chemical into the leaf.

---

[5] Of course, if the number of dark periods is the variable being studied, this would not be true, and in amateur studies more than one dark period will increase the flowering rate and often make the results easier to observe.

Time switches may control the light regime. Karl Hamner at the University of California at Los Angeles has reached some sort of an ultimate with a large number of modified file cabinets, kept in a temperature controlled room, which is provided with artificial lights of relatively high intensity. The file cabinets are modified by removing the upper drawers and replacing the front of the lower drawer so that the cabinet is light tight when the drawer is shut. The drawer is opened and closed by an electric motor controlled with a time switch. When the drawer opens, the plants are below the lights, and when the drawer shuts they are in the dark. With this arrangement Hamner has been able to study rather complex cycles of light and darkness (see Chapter 8), a tedious research problem if plants must be moved by hand in and out of darkrooms.

Often it is necessary to treat plants in the dark. This requires a safelight which will have no effect upon the photoperiodism response. Such a safelight should be of green color and very low intensity. To obtain pure green light we surround a green bulb (either incandescent or fluorescent) with two or three layers of green or blue-green cellophane. The late Robert Withrow (77) developed very special green safelights for use at the Smithsonian Institution in Washington D.C.

2. Light Interruption Studies

Short-day plants are inhibited in their flowering by interrupting the dark period with light, and long-day plants are promoted in their flowering by this same procedure. In experiments designed to study this effect the investigator needs a source of high intensity light with accurate time control, in an area close to the darkroom, protected so that moving the plants doesn't expose them to unwanted light. If the light interruption lasts for more than a very short time, leaf temperatures must be controlled by cooling. Withrow devised a refluxing system for removing infra-red wavelengths. The light passes through water causing it to heat and evaporate. This takes place within an enclosed container, and evaporated water is condensed on cold pipes and dripped back into the water.

In many studies, light quality must also be controlled, and often this may be done most conveniently by controlling the light source and using simple colored filters consisting of plastic or cellophane. In some installations, light is passed through diffraction gratings, prisms, or interference filters. At the plant industry station at

Beltsville, Maryland, a large carbon arc lamp is used as the source, and the light is broken into the various regions of the spectrum by passing it through a prism (see Chapter 7). This apparatus has provided considerable information about the response of plants to light. We are presently building a light flasher which will use a Xenon arc lamp, a set of shutters, water-cooled lenses, interference filters, and other items.

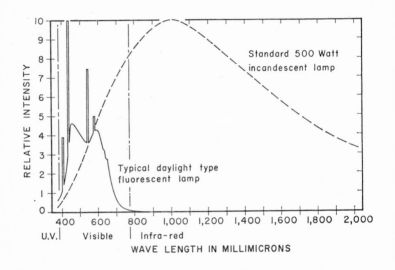

FIGURE 5–6

Spectral distribution of fluorescent and incandescent light. The vertical bars in the fluorescent lamp spectrum are emission lines of mercury. Data such as these are readily available from the lamp manufacturing companies (see also, for an excellent and thorough review of the problems concerned with lighting: Robert B. Withrow and Alice P. Withrow, 1956, Generation, control, and measurement of visible and near-visible radiant energy. In A. Hollaender (editor), *Radiation Biology* 3, 125–258. McGraw-Hill, New York).

The investigator often applies the following facts: incandescent light is rich in both red and far-red (far-red can be obtained by filtering through two layers of red cellophane and two layers of blue cellophane), while fluorescent light is rich in red but has essentially no far-red (see Fig. 5–6). Pure incandescent light acts primarily as a red source, even though it contains more far-red than red, because the pigment phytochrome (see Chapter 7) is more sensitive to the red than to the far-red. Thus it is quite possible to demonstrate the red-far-red phenomenon with only an incandescent source, a red cellophane filter, and a far-red filter.

3. *The Use of Chemicals*

Chemicals to be tested for their effects on the flowering process are usually dissolved in water to which a few drops of wetting agent is then added. Six drops per liter of liquid Tween-20 detergent will usually allow the solution to wet the leaves uniformly and give a good uptake of the compound in question. Other wetting agents will work equally well, and the amount used is not very important.

Often the investigator is faced with the rather serious problem of how to get the compound into solution. Acids and bases are often used, but one must avoid destruction of the compound, and the acidity (pH) of the solution will probably influence its effectiveness. Some compounds are soluble in alcohol, and fortunately cocklebur will tolerate solutions of 50 % ethyl alcohol with neither damage to the plants nor effects upon flowering. As standard practice, control plants are treated with solvent containing no chemicals.

In initial experiments with a new compound it is desirable to apply it over a range of concentrations. Usually a logarithmic concentration series of solutions is prepared. An aliquot of an initial solution is diluted to a predetermined volume. The same sized aliquot is then taken from the diluted solution and again diluted to the predetermined volume. We use series in which the numerical concentration values are repeated with a tenfold dilution at regular intervals in the series. The quantities to be used in preparing such series are listed in Table 5–2, and the results of an experiment of this type are shown in Fig. 5–7.

How should solutions be applied? Many investigators spray the solution on the plants. If the experiment is subsequently to be compared with experiments to be carried out in the field where chemicals are applied by spraying, this is probably a good method to use. Furthermore, certain plants such as rosette or cushion plants are difficult to treat with chemicals in any other way. But if the plant is caulescent (has a stem with leaves) such as a cocklebur, it is much easier to apply the chemical by simply dipping the leaf in the solution. This probably provides a more uniform application (the excess is

TABLE 5–2. DILUTION PROCEDURE FOR LOGARITHMIC
CONCENTRATION SERIES

	Quantity of original and successive solutions to be diluted to 1000 ml final volume[1]					
	100 ml	316.2 ml	464.2 ml	562.3[3]ml	631.0 ml	681.3 ml
	Concentrations resulting from successive dilutions:					
Original solution	.1000	.1000	.1000	.1000	.1000	.1000
First dilution	.01000	.03162	.04642	.05623	.06310	.06813
Second dilution	.001000	.01000	.02105	.03162	.03981	.04642
Third dilution	.0001	.003162	.01000	.01778	.02754	.03162
Fourth dilution	.00001	.001000	.004642	.01000	.01585	.02105
Fifth dilution	.000001	.0003162	.002105	.005623	.01000	.01468
Sixth dilution[2]	.0000001	.0003162	.001000	.003162	.006310	.01000

[1] Although 4-place figures are given, 3-place accuracy is usually sufficient (e.g. graduated cylinders may be used). Of course, the greater the number of dilutions in the series, the more the errors could multiply.

[2] Dilutions could be continued indefinitely.

[3] This series was used to obtain the results of Fig. 5–7. Note that concentrations appear at equal intervals on a logarithmic scale.

allowed to drip off), and problems of contaminating the atmosphere and surroundings are not so great.

There are many other special ways of applying solutions. To try to pinpoint the time of penetration of a chemical into the leaf, I have dipped the leaves in rather concentrated solutions and then washed them with distilled water after a measured interval of time. With a 5 min interval between dipping and washing, approximately 10 times the concentration was required for a given effectiveness as for dipping without washing.

Some chemicals are taken up by roots and not by leaves and must be applied to the soil. Cuttings of the plant may be placed in test tubes containing the solution, or the leaf blade may be removed and the cut end of the petiole allowed to absorb the solution from a small vial. In studies involving radioactive compounds, a single drop of solution may be placed on the leaf blade. In another approach, the leaf was immersed in a solution during an entire dark period, demonstrating that very low concentrations of certain chemicals (the auxins) would effectively inhibit flowering. These are cumbersome

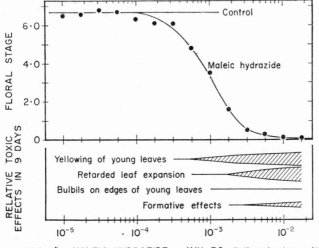

FIGURE 5–7

An example of an experiment in which chemicals are applied to plants in a logarithmic concentration series just before an inductive dark period. The control is expressed as a line or level, since there is no zero point on a logarithmic scale. Damage noted after 9 days is expressed by relative width of the wedge-shaped bars. The exact concentrations for this series are shown in Table 5–2. Floral Stages are explained near the end of this chapter. See F. B. Salisbury, 1957, *Plant Physiol.* 32, 600–608.

methods which should be used only when simpler ones fail. Since we know nothing about the concentration of a compound at its site of action inside of the cells, the concentrations given in descriptions of experiments must be considered in a relative sense and in light of the experimental procedures which were used.

4. *Other Methods and Techniques*

The methods described above are not a complete summary of those used in flowering studies. Since we are not going to consider vernalization in the remaining part of the book, methods relating to this problem are not mentioned here. Furthermore, the investigator will often devise special procedures to suit the particular problems at hand.

Grafting of one plant to another has been widely applied in translocation studies. One method is the approach graft. It is probably the simplest procedure which can be used to demonstrate

H

movement of the flowering stimulus from one plant to another. Epidermal and cortical tissue is removed from the stems of two plants, usually with a razor blade. These exposed areas are then pressed together, and the two stems are tied in place. Wide rubber bands may be used, although care should be taken not to wrap them too tightly. Special tapes and other materials are also available.

A more difficult procedure involves the grafting of a leaf from one plant onto another plant. The leaf is removed with a razor blade, usually taking some stem tissue along with the leaf and its petiole (in some cases it may be desirable to specifically exclude the bud in the axil of the leaf). This is then attached to a portion of stem from which the epidermal and some cortical tissues have been removed, as in approach grafting. Again the leaf is held in place by binding with rubber bands or special tape. The problem in leaf grafting is that the leaf tends to wilt. It will be 3 to 4 days before anatomical connections are made between the leaf and the stem so that the leaf can obtain sufficient moisture from the plant to which it is grafted. We found that we could successfully keep grafted leaves alive by putting them under a fine water spray or mist. If the spray system is turned on for a few seconds each minute, nearly all of the grafts are successful. Enclosing in plastic bags is another way of keeping grafted leaves or shoots from wilting.

5. *The Notebook*

The data listed in Table 5–3 should be recorded in connection with every experiment. Often interpretation makes it necessary to go back and check data which perhaps did not seem important at the time of recording.

IV. THE MEASUREMENT OF FLOWERING

1. *The Problem*

There are many steps between the vegetative plant and the mature flower, and this may strongly influence the way flowering is measured. The interests of the researcher will usually determine the method of measurement. If, as in the agricultural or horticultural sciences, one is primarily concerned with ultimate flowering, then the method of measurement is of relatively small consequence. The problem

TABLE 5–3. CHECK LIST FOR WRITING UP EXPERIMENTS

1. Experiment number and title. The title should be a brief description of the purpose of the experiment.
2. The date on which plants are given the first long dark period.
3. The size of the leaf left on trimmed plants.
4. The day on which plants are trimmed.
5. General weather conditions on the days preceding and following the long dark period(s), and greenhouse temperatures.
6. Number of days between the first inductive dark period and the time plants are dissected.
7. Age of the plants.
8. Name of people who helped with the experiment.
9. Experimental procedures including items such as:
   A. Length of the dark period(s), times of day when plants are put in or removed from darkness.
   B. Times of light interruptions, application of chemicals, etc.
   C. Concentrations of chemicals, mixing procedures, and wetting agents.
   D. Method of application of chemicals.
   E. Times, qualities, intensities, etc., of light interruptions.
   F. Any other pertinent data. Enough information should be given so that the experiment could be repeated exactly. Actual times of day should be listed, dark rooms used should be indicated, and no pertinent point should be omitted.
10. Flowering data: number of plants in each treatment with a given Floral Stage (see below), average Floral Stage and % flowering by treatments.
11. Graphical representation of the data.

becomes most acute when one is interested in the act of induction — those processes taking place within the plant up to the time when the shoot tip begins to develop into a flower. We have no way to study the act of induction except by observing ultimate flowering. Thus study of the act of induction is complicated by the intermediate development of the flower, which may be strongly influenced by many factors, including the overall rate of vegetative growth, which is in turn influenced by many environmental factors. This is an important and very significant complication in the overall question of how to study flowering.

2. *Some Methods Used by Various Workers* (3)

Does flowering occur in response to a given treatment or doesn't it? This is probably the simplest question that the investigator might ask. He can wait until plants have reached an advanced stage of maturity and record whether or not they have flowered. Very commonly some plants in a given treatment will flower and others will not, and results may be expressed as a percentage of flowering plants compared to the total replications of a treatment. A refinement of this approach is to record the percentage of plants which reach a certain flowering stage within some arbitrary time limit. Or the time required to reach a certain arbitrary stage may be recorded. It is common, for example, to present data as the number of days to the first visible bud. In such studies the rate of floral development is the factor actually measured, but it is often assumed that this is a function of the initial flowering stimulus produced by the act of induction. While this may be true, environmental effects on development should not be overlooked.

Another procedure closely related to these is to examine plants at some arbitrary time and assign arbitrary stages of floral development to these plants. Various workers have used both macroscopic and microscopic stages. A first problem arises in trying to assign the stages; a second arises in knowing what relationship the numbers assigned to the stages might bear to each other. In spite of these problems we use this method, and our reasons for adopting it are described in the next section.

Some workers have measured the number of nodes to the first flower, a convenient way of expressing data with certain species. The height of the flowering stem may be measured in plants that send up a flowering stalk from a rosette. The number of flowers or the number of flowering nodes may be counted. Such data may be compared to the number of leaves on the plant, allowing one to omit the time factor, which is always present in the methods described in the above paragraphs. William Hillman in his book (3) on flowering is quite concerned with this problem and suggests that we should always base our results on some sort of leaf index. In certain studies this can be quite important, and the expression of the data may have considerable influence upon the interpretation of the results. In my opinion, however, the time factor is seldom a complication when control plants are used correctly, and leaf index methods often involve waiting many weeks for the final data. If space is limited this will reduce the number of experiments which can be performed, and one must decide whether or not the advantages to be gained are worth the cost in terms of time and ultimate effort.

3. *Our Approach*

We are primarily interested in the act of induction, and thus would like to know the *amount* of flowering hormone produced, but our only measure of this amount is the appearance of the flower.

The rate of development of the cocklebur flower (specifically the male flower at the stem apex) seems to be primarily a function of the amount of flowering hormone which arrives at the tip. This conclusion, although a reasonable one, is probably based on a form of circular logic. We will discuss it further along with some examples in Chapter 9, but the basic idea is this: treatments which might be expected to influence the amount of flowering hormone reaching the tip also influence the subsequent rate of development of the bud.

We need a numerical measurement for this rate of development. This can be obtained by simply measuring the size of the buds at intervals after induction, and such a procedure has often been used. A better procedure is to establish a series of arbitrary stages of development of the bud. As a graduate student, I made about 80 photomicrographs of developing buds and then arranged the prints according to a logical sequence. The result was the series of 8 stages of floral development which are illustrated in Fig. 5–8 and defined in Table 5–4. The stages involve morphogenetic changes, and these seem to be less influenced by the plant's general health than is the

TABLE 5–4. CRITERIA FOR THE DETERMINATION OF FLOWERING STAGES
IN THE APICAL BUD (TO BECOME STAMINATE) OF
Xanthium pennsylvanicum, COCKLEBUR

Floral stage	Criterion
0	Vegetative. Shoot apex relatively flat and small.
1	First clearly visible swelling of the shoot apex.
2	Floral apex at least as high as broad, but not yet constricted at the base.
3	Floral apex constricted at the base, but no flower primordia yet visible.
4	First visible flower primordia, covering up to the lower one-quarter of the floral apex.
5	Flower primordia covering from one to three-quarters of the floral apex.
6	Flower primordia covering all but the upper tip of the floral apex.
7	Floral apex completely covered by flower primordia. Slightly to moderately pubescent.
8	Very pubescent and showing some differentiation of flower parts. At least one millimeter basal diameter.

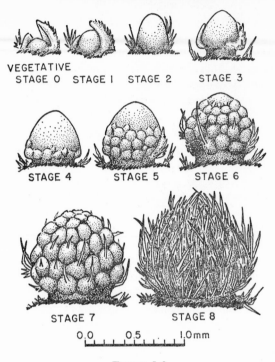

VEGETATIVE
STAGE O   STAGE I   STAGE 2   STAGE 3

STAGE 4   STAGE 5   STAGE 6

STAGE 7   STAGE 8

0.0   0.5   1.0 mm

FIGURE 5–8
Drawings of the developing terminal inflorescence primordium (staminate) of cocklebur, illustrating the system of flowering stages described in Table 5–4. See F. B. Salisbury, 1955, *Plant Physiol.* 30, 327–334.

size of the bud. The stages are easy to recognize, and buds can be classified much more rapidly than their size can be measured. Of course it is sometimes difficult to classify a particular bud when it is at a stage midway between two of the defined ones. In such cases the plant is arbitrarily assigned to a higher stage, and the next time such a case arises within the same treatment, the lower stage is assigned so that the errors will tend to cancel out when the average flowering stage[6] is determined for a given treatment. Half stages can also be used. Averaging minimizes the errors: if there are 10 plants in a treatment and an error of one stage is made in classifying a bud, there will be an error of 0.1 stage in the Floral Stage. This is quite

[6] The term Floral Stage as used here is meant to imply the *average* flowering stage for the replications within a treatment.

small and should not concern us much at this point in the development of the science.

The numerical values assigned to the 8 stages happen to be such that when one expresses Floral Stage as a function of time after the dark period, the points approximate nearly straight lines, as shown in Fig. 5–9. Development usually begins about $2\frac{1}{2}$ days after the

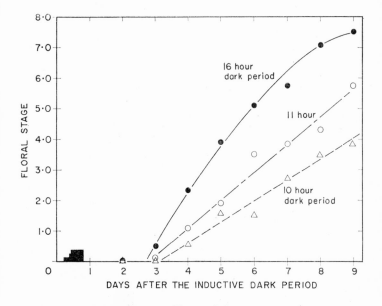

FIGURE 5–9

Floral development curves; Floral Stage as a function of time after induction by a 10, 11, or a 16-hr dark period. Some of the points lie unusually far from the lines for this kind of experiment, but as might be expected (Fig. 3–7) this is most true for the shorter dark periods. Inductive dark periods were initiated on June 12, 1962. Data previously unpublished.

inductive dark period, and the line continues fairly straight up to 9 to 12 days, when controls have usually reached the maximum stage. Since the lines are nearly straight (or at least curve in about the same way for various treatments), it is not necessary to measure flowering by following development rate in this way. Stages reached by some arbitrary time after the inductive dark period will be proportional to rate of development of the buds during this time. In our standard procedure we examine the buds under a binocular, dissecting micro-scope (about 36 diameters magnification) 9 days after the beginning of the experiment. This is quite arbitrary, but it is convenient.

What about environmental effects upon rate of floral development? Under certain conditions such effects may be very striking. For ex-ample, with an extremely low night temperature (only a few degrees above freezing) and a relatively cool day temperature, no development occurred at all as compared to a very rapid development when temperatures were high (Fig. 5-3). Yet in practice, if conditions will allow a fair development of induced control plants, the problem is not a serious one. The control plants are subjected to the same developmental environment as the treated plants, which is insured by the way that they are placed on the greenhouse benches. Thus only the treatments vary, and everything else is maintained com-parable between controls and treatments. Floral Stages of treated plants are not considered in an absolute sense, but only as they compare with untreated controls in the same experiment. This is the procedure which is commonly followed in science, and indeed it might be used to study floral development itself (as in Fig. 5-3).

It might be even better to standardize environmental conditions in such a way that the flowering controls always develop to the same stage in the same number of days, providing that they have been given the same inductive treatment. Temperature, light conditions, fertilizer and moisture could all be standardized to realize this end. When this is achieved, we will logically make absolute comparisons between experiments, as well as between treated and control plants within an experiment. Such a condition has yet to be reached, but it is being approached at present, especially in growth chamber experiments.

CHAPTER 6

PREPARATION FOR RESPONSE TO PHOTOPERIOD

THE higher plants will not flower, nor in many cases respond to the environmental stimulus which leads to flowering, until they have reached some minimal age; that is until they have reached "ripeness to flower". The flowering machine must be constructed. Furthermore, experiments showing the necessity for high intensity light (probably photosynthesis) preceding the inductive dark period seem to indicate that an energy source is essential before the machinery of flowering will begin to function. Ripeness to flower and the high intensity light process, then, will be discussed in this short chapter.

JUVENILITY AND RIPENESS TO FLOWER
(3, 13, 20, 24, 31, 37, 72, 76)

Probably all of the common species used in studies on photoperiodism and vernalization have been investigated as to the minimal age at which they will respond to the environmental stimulus. As mentioned in Chapter 4, the vernalization requiring cereals will respond even before the embryo has reached maturity in the developing seed on the mother plant. The later requirement for long days, however, can be satisfied only after they have produced a few leaves which will respond to the long-day stimulus. The biennial strain of *Hyoscyamus niger* must be 10 to 30 days old before it becomes sensitive to vernalization. *Chenopodium rubrum* will respond to short days and flower as a seedling on moist filter paper in a petri dish. The cotyledons of Japanese morning glory respond to short days, but the cocklebur is probably more typical, in that the cotyledons and very young leaves (less than a centimeter in length) will not respond, but true leaves longer than about 2 cm are sensitive (see Fig. 5–5). On the other end of the scale are many trees which will not flower until they are 5 to 40 years old, and certain bamboo species grow for 5 to 50 years or longer, flower once, and then die. Presumably the environmental stimulus to flower, if one is required, is received each year, but the machinery to respond to it is not developed until the plant has reached ripeness to flower. Often it must grow beyond a so-called juvenile stage to one of relative maturity.

The juvenile stage is itself an interesting topic for discussion (76). Juvenile features are often exhibited in many respects besides the inability to flower in response to environment, and often this feature is not well correlated with the others. Leaf shape of the first leaves may differ markedly from the shape of later leaves. This is fairly well illustrated in Fig. 1–1–B for the cocklebur plant (here the juvenile leaves have already reached ripeness to flower). Arrangement of leaves on the stem (phyllotaxis) may also change during maturation. The first two cocklebur leaves are usually opposite, while successive leaves become more and more alternate. There are often changes in the shape of thorns, changes in ability to form sun or shade leaves or to form roots on cuttings, and changes in other features. Although the change from the juvenile to the mature condition is usually gradual, the two forms may be so different from each other that they can easily be mistaken for different species. The two forms of ivy, one a vine with palmate leaves and the other a shrub with entire leaves, are actually treated as separate horticultural varieties.

A striking feature of the change from the juvenile to the mature stage is the resulting stability. Cuttings from adult parts of the plant (the base of a tree will often retain its juvenile characters) maintain their adult characters when they are rooted or grafted to young root stocks. A famous example is a cutting made from a plagiotropic (horizontally growing) adult branch of *Araucaria excelsa* which has retained its horizontal growth habit for more than 50 years at the Munich Botanic Garden. It appears that some basic change has taken place in the tissue, and that this change is preserved through cell divisions as the plant continues to grow.

There is some evidence that hormones are involved in the manifestations of juvenility. Perhaps most striking is the reversion of adult ivy to the juvenile stage after treatment with gibberellins (73).

Other changes take place in growing plants which are not to be confused with maturation, but which seem to be part of the ageing process. It is well known that most organisms exhibit a sigmoid growth curve (see Chapter 9 for examples with cocklebur), in which there is an initial rapid period of logarithmic growth followed by a decline in growth rate. Often the decline in growth rate is accompanied by changes in the plant's morphology, such as an increased number of spur branches on certain trees. This ageing process can be distinguished from maturation, because cuttings taken from old trees, which are rooted or grafted to young root stocks, regain the growth characteristics of youth. There is good preliminary evidence that ageing is mostly a matter of nutrition, although hormonal factors might also be involved. At any rate, there is some reason to think that ageing and flowering might be related, since treatments of trees (mutilations, etc.) which will hasten ageing may also tend to hasten flowering.

Whatever the relationship might be between maturation, ageing, and ripeness to flower, there are three specific questions which can be asked in relation to flowering: First, which part of the plant must attain ripeness to respond? In photoperiodism we might ask whether the leaf must become mature enough to respond to day-length or the meristem must become mature enough to respond to the flowering hormone. Second, if a plant requires more than one season to attain ripeness to flower, is it because the plant must be exposed to a given number of cycles in activity and dormancy or because it must simply grow for a given period of time? Are the cycles counted or the active days added together to reach a total? Third, what is the physiological nature of ripeness to flower?

1. *Which Part of the Plant Must Attain Ripeness?*

Figure 5–4 in the last chapter showed that the sensitivity of the various leaves on a cocklebur changes as the leaves mature. Sensitivity of a leaf is dependent upon its size (rate of growth — see Chapter 9) and not age of the plant. In Perilla, however, it is possible to show by inducing detached leaves that the ones produced by mature plants are more sensitive than the ones produced by very young plants. Both observations seem to imply that ripeness to flower must be a matter of the leaf reaching maximum sensitivity to day-length, but we cannot exclude the idea that the meristem must also reach some sort of maturity.

J. A. D. Zeevaart (80) was able to perform an experiment which has direct bearing on the problem. He took very small *Bryophyllum* plantlets, which will not respond to photoperiod by flowering and which have a long juvenile phase, and grafted them onto large flowering plants. The small plantlets began to flower immediately. Thus the meristems were capable of responding to the hormone but the leaves on the seedlings apparently cannot respond to the environment by producing hormone.

Similar results have been obtained with a few other herbaceous species, but as implied above, grafting of juvenile woody shoots onto mature trees will not induce immediate flowering of the juvenile shoots.

2. *Is Ripeness Attained by Counting Cycles or by Summating Total Active Growth Time?*

K. A. Longman and P. F. Wareing in Wales (60) set up an experiment in 1957 designed to solve this problem for birch (*Betula verrucosa*) which requires 5 to 10 seasons to flower in nature. One group of seedlings was grown continuously under 18-hr days in the greenhouse; another group was given long days until 25 to 30 cm of growth had occurred, then short days (9 hr), causing the plants to become dormant, 6 weeks of chilling to break dormancy and then long days again, and so on. After 10 months the seedlings under continuous long days began to flower when they were 2 to 3 m tall. Within two years, 13 of the original 14 trees had flowered. None of the other trees given 7 short, simulated seasons had flowered by then, even though that many full seasons in nature would have probably caused flowering. Thus ripeness to flower in this instance seems to depend upon the total growth attained rather than the number of cycles of growth and dormancy.

3. *What is the Physiological Nature of Ripeness to Flower* (37)?

It is known that sugar is necessary for vernalization of winter rye seedlings (see Chapter 4), and it has been suggested that ripeness to flower is a matter of building up enough reserve carbohydrate. Furthermore, the juvenile phase can be shortened in *Lunaria biennis* (a biennial) by strong additional light which may increase photosynthesis and reserve material. In other biennials vernalization does not take place after defoliation unless the plants have reserve organs such as tubers. Isolated apical buds of carrot and beet can be vernalized, but only in the presence of sugar solutions. Only one observation seems inconsistent with the reserve foodstuff viewpoint of ripeness for vernalization. One might imagine that the ability of a moist seed to be vernalized would be related to the amount of reserve materials it contains, that is to its size. Such a relationship does not exist. Some very small seeds can be vernalized and some very large ones cannot. If it is a matter of reserve materials, quality would seem to be more important than quantity.

We have no specific information relating to the nature of ripeness to flower in leaves of photoperiodically sensitive plants. We can, however, probably state in principle that ripeness to flower must be a matter of getting the machinery for response to photoperiod and/or synthesis of flowering hormone constructed and in operating order. The right precursors must be available, and the right enzymes must be on hand; the pigment system (which is probably present all the time) must be coupled to the clock and to the florigen producing system in the proper way. Getting any or all of these processes into functioning condition could be ripeness to flower. The fact that it occurs in a specific organ of the plant, the leaf in photoperiodism or the meristem in vernalization, is another example of the basic biological question of how certain genes (the ones for producing flowering machinery) can operate in certain places and not everywhere in the plant.

THE HIGH INTENSITY LIGHT PROCESS

It was known that response to an inductive dark period in the case of short-day plants was best if this inductive dark period was both preceded and followed by light of high intensity. Dim light either before or after (see Chapter 10) the dark period caused a reduction in the degree of flowering. In 1940, Karl Hamner, then at Chicago, clearly demonstrated the necessity for high intensity light before the inductive dark period. He treated plants with short dark periods separated from each other by brief intervals of light. The short dark periods were not long enough to allow synthesis of flowering hormone, nor were the interruptions with light bright enough to allow a significant amount of photosynthesis. If a long dark period followed this treatment immediately it was ineffective, and plants remained vegetative, but if plants were exposed to 1 to 3 hr of sunlight after the short dark periods, then the subsequent long dark period was effective and the plants flowered. This is illustrated in Fig. 6–1.

The experiment is easily interpreted assuming that effectiveness of the long dark period depends upon the presence of sugars or other energy sources which are produced by photosynthesis and used up

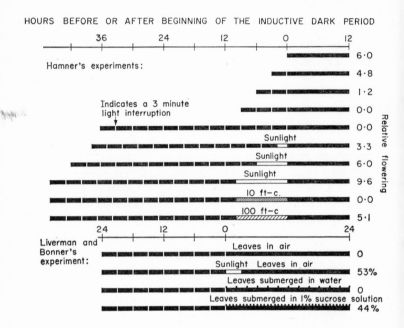

FIGURE 6–1

Some of the experiments of K. C. Hamner (*Botan. Gaz.* 101, 658–687, 1940) showing the effects on flowering of cocklebur of intermittent light and dark periods and other treatments preceding a dark period of normally inductive length; and experiments of J. Liverman and J. Bonner (*Botan. Gaz.* 115, 121–128, 1953) showing effects of sugar applied during a long dark period which has been preceded by intermittent light and dark periods. It was possible to convert Hamner's flowering data roughly to the Floral Stage system described in Chapter 5. Liverman and Bonner present only data for the per cent of the treated plants which flowered. Before and after treatment, plants were subjected to long days in the greenhouse. Buds were examined after 3 weeks. Solutions were applied by immersing the leaf in the solution during the entire inductive dark period. In other experiments solutions were applied by putting cuttings into the solution at the beginning of the long dark period. Three minute light interruptions were 200 ft-c (Hamner) or 50 ft-c (Liverman and Bonner).

during the flashing light treatment. James Liverman and James Bonner at the California Institute of Technology provided supporting evidence for this idea. They were able to make a long dark period preceded by short dark periods effective by treating plants with various sugars such as might be produced in photosynthesis. Anton Lang (55), then at Los Angeles, has also successfully used reducing substances such as ascorbic acid and glutathione. The interpretation of these experiments is not completely straightforward, because the reducing substances might themselves be used as energy sources, or the sugars might be used to produce reducing substances. An experiment will have to be devised to see whether flowering specifically requires reducing substances or just any energy source. It does seem clear that energy is required for operation of the flowering machine, since respiration inhibitors will nullify an otherwise effective dark period.

Specifically, production of ATP through respiration seems to be essential to effective action of the inductive dark period, since treatment with dinitrophenol makes this dark period ineffective (see Chapter 9). It has also been shown that photosynthesis is an important part of the flowering process in long-day plants.

Actually, the experiments described above are not as easy to interpret now as they were a number of years ago. We shall see in Chapter 8 that a brief light flash such as those effective in separating the short dark periods in Hamner's experiment may not reset the timing mechanism of the flowering process, while a longer exposure to high intensity light might so reset the flowering clock. Thus Hamner's original experiment could also be interpreted on the basis of the timing considerations which are discussed in Chapter 8. The short flashes might inhibit flowering to a degree determined only by the timing mechanism, and the long exposure to sunlight might reset this timing mechanism. But how can we relate this to the results with sugar and reducing substances, which seem to explain everything on the basis of photosynthesis and respiration? What relationship can sugar and reducing substances have to the timing mechanism? Some new experiments will have to be performed to straighten out some of these problems.

In spite of these difficulties, it seems amply clear (based on the respiration inhibitor evidence at least) that the flowering machine must have an energy source for effective operation.

CHAPTER 7

LIGHT AND THE PIGMENT

Two paramount aspects of the flowering process are important in any photoperiodic response, and they form the roots from which the term photoperiodism was derived: response to light and the measurement of time. This chapter discusses what is known about the way that the plant senses when the lights are on or off, and the next chapter considers the manner in which the plant might measure the length of light and dark periods.

THE PIGMENT SYSTEM IN FLOWERING

1. *Pigments in General* (8)

A basic rule of photochemistry is that light must be absorbed in order to be effective in any process involving chemical change. If the light is not absorbed, no response can result. If a molecule absorbs light, then it will appear to our eyes to be either colored or black, and we refer to it as a pigment. If all visible wavelengths are absorbed, it appears to our eye to be black. If any wavelength in the visible spectrum is not absorbed, then the compound will appear to be that color. If the red, yellow, and blue wavelengths of light are absorbed and green wavelengths are not, then the compound appears green (e.g. chlorophyll) but if green is absorbed and blue and red are left, then the mixture is sensed as purple. If only blue is absorbed the remaining wavelengths give us the impression of yellow. The pigment to be discussed in this chapter absorbs mostly red and thus appears to be bluish-green.

In the flowering process, there is a response to light, meaning that the light must be absorbed by a pigment. Finding and characterizing the pigment involved in photoperiodism would be an important initial step in understanding the process. Knowing the color should help us find the pigment, and knowing the wavelengths absorbed should tell us the color. The approach is to measure the relative effectiveness of various wavelengths in the process at hand, providing a so-called action spectrum for that process. If only the effective wavelengths are absorbed by the pigment, as seems fairly logical from the first law of photochemistry, then the color properties of the pigment can be described.

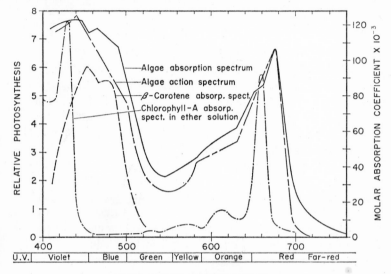

WAVE LENGTH IN MILLIMICRONS

FIGURE 7-1

The action spectrum for photosynthesis in *Ulva taeniata*, a green alga, compared with the absorption spectra of the alga, chlorophyll-α, and beta-carotene. The action spectrum most closely matches the absorption spectrum of the alga, except for a shoulder between 460 and 500mμ which is accounted for by the photosynthetically inactive carotene. The absorption spectrum for chlorophyll probably accounts for most of the absorption spectrum of the alga and the action spectrum, but in ether solution the peaks are very sharp and shifted to the left compared to the spectrum produced by chlorophyll in water or the cell. Data for the isolated pigments from various plant physiology texts; for the algae from F. T. Haxo and L. R. Blinks, 1950, *J. Gen. Physiol.* 330, 389–422.

The classical example is that of photosynthesis, and the action spectrum for the process in a green alga is compared in Fig. 7-1 to the absorption spectra of chlorophyll, carotene, and the plant as a whole. Obviously the match between absorption of chlorophyll in ether solution and the action spectrum for photosynthesis is far from

I

perfect. The absorption of the whole plant is much closer to the photosynthesis action spectrum, but one shoulder in the plant curve at 460 to 500 millimicrons does not match the photosynthesis curve but is accounted for by the carotene curve. The many problems in this approach are well illustrated (e.g. the absorption spectrum of chlorophyll in ether solution is not the same as the absorption spectrum of chlorophyll in the cell).

2. *Method of Studying the Pigment in the Flowering Process* (19)

Properties of the photoperiodism pigment were studied beginning about 1940 by workers at the United States Department of Agriculture Plant Industry Station at Beltsville, Maryland. A piece of equipment essential to the study of this problem was one which would produce light of very specific wavelengths at relatively high intensities. The problem was primarily to obtain a source of white light of an intensity which would provide ample light in relatively narrow wavelength bands after passage through a prism. Such a light source was obtained from the rejected first model of a movie projector which had been constructed for one of the world's largest theatres. Light from the powerful arc could be focused with reflectors and then passed through condensing lenses and the prism to produce an intense spectrum, even when the spectrum was 3 m wide from the blue to the red end. Of course, intensity could be controlled in a number of ways, such as by controlling the aperture (or slit) through which the beam of white light passed or by adjusting the distance of the test plants from the light source.

The actual measurements were made by exposing plants to one or more long dark periods, which would normally promote flowering of short-day plants or inhibit flowering of long-day plants, and then interrupting each of these dark periods by placing the leaves of the test plants in various regions of the spectrum at different intensities for different lengths of time. The time-intensity combination which allowed about 50% of the plants in a treatment to flower was determined. Obviously a great many experiments were required. Groups of plants had to be treated in such a way that each plant in the group received the same time, intensity, and wavelength of light. These plants then had to be examined a number of days later, and often the experiment had to be repeated, making appropriate adjustment in time and intensity to obtain the 50% flowering level. Of course various combinations were tried in each experiment, but the work had to be repeated many times before the action spectra shown in Fig. 7–2 could be obtained, in which the quantity of light required to inhibit flowering 50% is plotted as a function of the wavelength.

If such curves are inverted so that the least required energy is at the top, as in Fig. 7–3, they are comparable to the absorption curve of a pigment in which the amount of light absorbed is greatest at the top (also shown for the photoperiodism pigment in Fig. 7–3). That is, the wavelength which is effective at lowest intensities must be the wavelength which is absorbed most by the pigment.

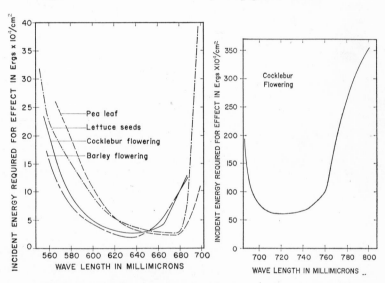

FIGURE 7–2

Action spectra expressed as light energy required for half effect in a number of systems. Red effects are for 50% promotion of germination of lettuce seeds, 50% inhibition of flowering in cocklebur, 50% enhancement of dark grown pea leaf elongation, and 50% promotion of flowering in barley. The differences in the peaks are probably accounted for by chlorophyll screening. The far-red reversal curve is a generalized one for a number of systems. Original data are available in many papers by the Beltsville workers (see, for example, 8, 10, 11, 18 and 19).

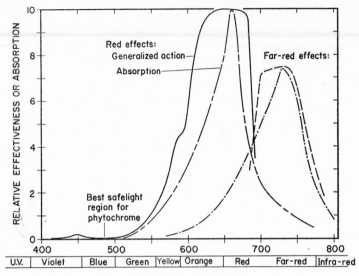

FIGURE 7–3

Generalized action spectra expressed on a scale of relative effectiveness, and absorption spectra of the two forms of phytochrome as measured primarily *in vitro*. The broad peaks for the action spectra are meant to imply that various plant systems might have sharper peaks within the indicated broad range. Difference in such sharp peaks may be ascribed to side effects such as chlorophyll screening. Absorption data from the Australian symposium paper of Hendricks and Borthwick (19).

3. *Control by the Pigment of Responses Other Than Flowering*
(7, 8, 10, 18)

A striking thing about the Beltsville experiments is that the action spectra (Fig. 7–2) for short-day plants (an inhibition of flowering) is very similar to the action spectra for long-day plants (a promotion). Orange-red light is most effective in either case. Actually, by the time the curve was obtained, it was already possible to recognize its familiar shape. It is evident in Fig. 7–2 that the flowering curves are very similar to curves showing the promotion by light of lettuce seed germination, and of dark grown pea leaf elongation, which were obtained in much the same way with the same instrument. The curves also closely resemble those for the inhibition of stem elongation in the dark (etiolation), the straightening of a hook at the end of a bean stem, and other plant responses. The immediately obvious conclusion has fairly well withstood the test of time during the past 15 or 20 years: the pigment absorbing the light which is effective in the flowering process is the same pigment which is effective in germination, halting of etiolation, plumular hook straightening, and many other responses. This is the kind of generalization which delights the heart of any scientist, the kind which is always hoped for, but seldom encountered, especially in biology.

4. *The Reversible Nature of Phytochrome*

At this point in the unfolding of things (about 1951) the group at Beltsville happened to have one of the truly lucky breaks which sometimes reorient an entire research program — or even the research in a major area of study such as plant physiology. It was known that lettuce seeds placed in the spectrum were not only promoted in their germination by orange-red light, but that they were somewhat inhibited by light of still longer wavelengths, so-called far-red. Someone then had a rare insight: the effects of red light might be reversed by a subsequent illumination with far-red light. This guess turned out to be correct! If lettuce seeds were illuminated with red light, nearly all of them germinated, but if seeds which had been so illuminated were then placed in the far-red part of the spectrum, the promoting effects of the red light were overcome, and actually fewer seeds germinated than in a control treatment left in the dark. Furthermore, if they were exposed to red and then far-red and then red, they would germinate. If red and far-red were alternated many times, seeds would germinate if the last illumination period had been red, or they would not germinate if it had been far-red, as shown in Fig. 7–4. The action spectrum for far-red reversal of the red effect is shown in Fig. 7–2 and 7–3 along with the red action spectrum.

A theory developed immediately: the plant must contain a pigment which changes its form by absorbing red light, and the form it becomes is one which will absorb far-red light. When the far-red absorbing form is illuminated with far-red, it changes back to the form which absorbs red. Of course the crux of the theory was that the biological response of the plant depended upon the form of the pigments. After the pigment's extraction (see below), it was called phytochrome. The reactions are summarized in the following formula:

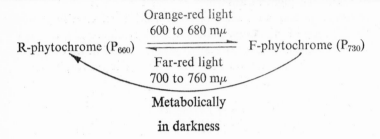

R-phytochrome (P$_{660}$)   Orange-red light 600 to 680 mμ / Far-red light 700 to 760 mμ F-phytochrome (P$_{730}$)

Metabolically

in darkness

When the pigment is in the R-phytochrome form, the plant's biochemistry, at least as it relates to flowering, is adjusted in some way so that it is typical of a plant in the dark. Changing the pigment to F-phytochrome induces biochemical changes which inhibit flowering of short-day plants and promote flowering of long-day plants. An essential part of the formula is the shift from F-phytochrome to R-phytochrome in the dark (with no illumination by far-red light). This shift fails to occur with extracted phytochrome in the test tube, so it must depend upon the plant's metabolism.

5. *Extraction of Phytochrome* (10, 19)

Information about the reversible nature of the pigment should be of considerable value in searching for it. One would look not only for a greenish-blue pigment with a certain absorption spectrum, but for a pigment which changes color (absorption spectrum) when it is illuminated with red light and then changes back again when it is illuminated with far-red light. With this information, researchers initiated an intensive program aimed at isolation of the pigment from plant tissue. It was ultimately the Beltsville group that succeeded. At least 7 years were spent before any degree of success was obtained, and the pigment has yet to be completely isolated and characterized, although its presence in a test tube can now be clearly demonstrated.

An immediate problem concerned the source of the pigment. If one were to use cocklebur leaves, for example, the isolation of one greenish-blue pigment present in very small quantities would be extremely difficult in the presence of chlorophyll, a bluish-green pigment present in very large quantities. So the approach was to use

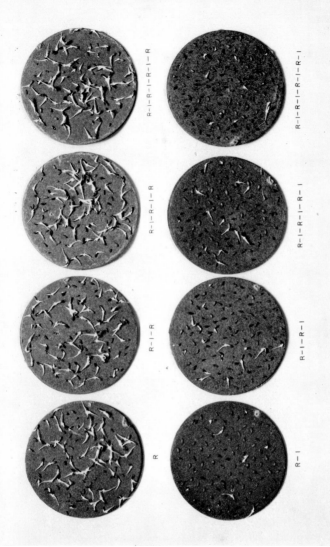

Reversal of lettuce seed germination with red and far-red light. Legends under individual lots indicate sequence and number of red and far-red treatments. Red irradiations were for 1 min and far-red irradiations for 4 min. If the last exposure is to red, seeds germinate; if to far-red, they remain dormant. Temperature during the half hour required to complete the treatments was 7°C. During all other times temperature was 19°C. Photograph furnished by Harry Borthwick, Agricultural Research Service, Plant Industry Station, United States Department of Agriculture, Beltsville, Maryland.

a tissue which was more or less colorless. Corn seedlings do not synthesize chlorophyll when they are grown in the dark but do respond to red and far-red light. Such organisms should provide an excellent source of plant material for isolation of phytochrome. Obviously, since the seedlings appear colorless, the pigment must be very dilute.

Since the pigment is so dilute, the primary problem is to be able to measure it with sensitive instruments. The spectrophotometer which finally matched the requirements was originally built to measure very slight changes in the light intensity of a light beam passed through a very dense sample of material. Intensity changes amounting to a small fraction of 1% could be measured in a beam of light passed through 5 cm (2 in.) of wood! Thus a great many corn seedlings could be packed together in a relatively large container, and light of any wavelength which was absorbed by these seedlings could be measured with a high degree of accuracy.

Next came the problem of learning how to recognize the pigment with this instrument. The solution was to illuminate the sample with high intensity red or far-red light and then immediately to measure the difference in absorption of red and far-red. The source of red and far-red for measuring the absorption alternated about 16 times each second. Pre-illumination with red should result in high absorption of far-red compared to red and vice-versa.

Early tests with the corn seedlings demonstrated that phytochrome could be recognized with the instrument. Then corn seedling tissue could be fractionated in various ways, and the most active fractions could be determined. Following this procedure, phytochrome, which proved to be a protein with a molecular weight of about 200,000, was extracted first in a concentration of about 10^{-7} moles per liter. It has been concentrated and purified by special column chromatography to the point where it is about 3% pure in a pellet of protein. In this form and in water suspension it is possible to see the bluish-green color and the slight color change that occurs upon illumination with red followed by far-red light. By the summer of 1962 the pigment had also been extracted from green plants. It is interesting that pigment could be obtained from a number of long-day plants but not from the short-day plants which were tried, including cocklebur. It is obviously also present in short-day plants, and the difficulty in obtaining it could prove to be a clue to its action in flowering.

6. The Nature of Phytochrome

The protein nature of the pigment is especially interesting. Physiologists have known for many years that some very important pigments are attached to protein. Chlorophyll is apparently attached to a protein when it absorbs the light used in photosynthesis. When chlorophyll is extracted, however, the protein is left behind, and only the relatively small colored molecule goes into solution. The red colored heme is also easily separated from the protein globin of the hemoglobin in blood. The phycobilin pigments of the red and blue-green algae also absorb light used in photosynthesis. These pigments are so tightly attached to protein that removal of them results in a drastic change in their properties, including the color. In the case of chlorophyll and hemoglobin, molecular structures of the relatively small, colored molecules have been determined. This has never been successfully accomplished with the phycobilin pigments, and it appears that it will not be an easy task with phytochrome. Studies so far have failed to separate the colored portion of the phytochrome molecule from the protein; indeed we might anticipate that it is attached even more strongly than the chromatophores of the phycobilin molecules which it resembles closely in absorption spectrum (e.g. allo-phycocyanin). At any rate the fact that phytochrome is a protein is encouraging, because this could well imply that it has enzymatic activity. So the next problem, even if we are unable to further determine the structure of phytochrome, might be to determine its enzymatic role in photoperiodism.

The Beltsville workers have now had a chance to study a number of phytochrome's properties by studying the isolated phytochrome. Work with whole plants had shown that the light driven conversions were independent of temperature, for example, and it has been shown that photo-conversion in the test tube will take place at temperatures as low as $-78°C$ (but not below). The pigment is rather easily denatured by temperatures of $50°C$, repeated freezing, or simply allowing it to stand at room temperature for several hours. Color changes occur upon denaturation, but the dark shift so essential to phytochrome's control of physiological processes does not occur in the test tube.

PHYTOCHROME AND OTHER PHOTOMORPHOGENIC RESPONSES

As mentioned above, the action of phytochrome is very general in the plant kingdom. Table 7–1 contains the examples which have

TABLE 7–1. PLANT GROWTH RESPONSES UNDER THE CONTROL
OF THE PHYTOCHROME SYSTEM

Most of the following responses are listed in 8, 10, 18, 19 or 36. Some of the less well-known examples, or some not listed in these reviews, have references cited in the table.

1. Time-independent responses, the degree of response usually related to the amount of F-phytochrome produced by irradiation.
   A. Elongation or enlargement responses.
      1. Stem elongation of vascular plants.
      2. Petiole elongation.
      3. Root growth (54,71).
      4. Leaf enlargement.
      5. Plumular hook unfolding (for interesting response in lettuce, see 65).
   B. Pigment formation.
      1. Anthocyanin formation in various systems such as apple or turnip skin, cabbage leaves, etc.
      2. Carotene formation in tomato skins.
      3. Activation of the chlorophyll synthesizing mechanism.
   C. Process initiation.
      1. Germination of many seeds and some spores (e.g. in moss, fern, see 39 and 65).
      2. Growth in the dark on a sucrose medium (heterotrophic growth) of *Lemna minor* or of fern gametophyte (64).
      3. Turning of chloroplasts towards the light in the green alga *Mougeotia* (50).
      4. Change to the gametophyte in a sporophyte culture of a moss (*Psycomitrium pinforme*) in response to 10 minutes far-red light before darkness (unpublished data of L. Bauer, Botanisches Institut, Tübingen).
      5. Initiation of timing in certain circadian rhythms (61).
II. Time-dependent responses: photoperiodism. The degree of response may be highly quantitative but also time dependent.
   1. Flower initiation in long-day and short-day plants.
   2. Normal and abnormal floral development in many species (e.g. sex determination in hemp, 44; cleistogamy, in which flowers fail to open and complete development, 43; and phylloidy of bracts in which bracts increase in size and resemble foliage leaves if floral induction is minimal, 49).
   3. Various morphological responses of the vegetative plant to daylength (e.g. succulency of Crassulacea, tuber or bulb formation, and inhibition of axillary buds in cocklebur, 41).
   4. Development of various reproductive structures in bryophytes (40).
   5. Germination of certain seeds.
   6. Onset of dormancy in many plants, sometimes breaking of dormancy.
   7. Carbon dioxide fixation in succulents.

come to my attention. Any degree of certainty that the process is really controlled by phytochrome should rest at least upon evidence that red light strongly influences the process, while far-red light reverses this influence (or in some cases vice versa). All of the time-independent responses in the table have been studied in this way, although some of them have been studied much more intensively than others. In the case of photoperiodism, however, some responses are included simply because photoperiodism itself has been clearly shown in a few instances to be a phytochrome-controlled response (e.g. flowering, germination of seeds, and inhibition of axillary buds in cocklebur). In other cases evidence with controlled light qualities is either not available or conflicting. Thus certain aspects of floral development, as well as development of reproductive structures in bryophytes, have been shown to be controlled by photoperiod, but no work has been done to show that red or far-red light is effective in promoting or inhibiting the process. In the case of dormancy, J. P. Nitsch (29) has preliminary results indicating that phytochrome takes part in the response, but P. F. Wareing obtained negative results with woody species in a preliminary experiment which was to be reported in this book. We have also obtained negative results with some alpine species. Thus much work remains to be done with plant responses to photoperiod.

It is interesting to note that the table contains some good examples of phytochrome control in the green algae, the bryophytes, the ferns, and in the higher plants including both gymnosperms and angiosperms. No evidence has been found for phytochrome in the fungi nor in the algae other than the green algae, however. Thus phytochrome appears in the evolutionary line in which chlorophylls A and B are the only light receptors in photosynthesis. Phytochrome may be absent from the other lines, but of course future work could change this viewpoint.

Is it not rather impressive that so many natural plant processes are under the control of a single pigment system? It is an evidence for parsimony and uniformity in natural processes. At the same time it poses an intriguing question: How can a single biochemical mechanism control so many different kinds of processes? The workers at Beltsville have suggested that some very central compound in metabolism might be involved, such as coenzyme-A. Reactions of coenzyme-A are certainly general; it takes place in carbohydrate, fat, and protein metabolism as well as in some other reactions. There are other biochemical reasons for suspecting this reaction, most of which come from some of the best work which has so far been done with the phytochrome system. The synthesis of anthocyanin pigments in apple skins, turnip skins, red cabbage leaves, and other plant materials has been studied intensively by the Beltsville workers. Such a process offers an excellent opportunity for the study of biochemical mechanisms, since we already know something about the biochemical steps involved.

Other Photomorphogenic Pigment Systems

It is quite easy to become so impressed with studies on phytochrome that one overlooks the possibility that other important pigment systems might exist, which could also control growth. Of course it is well known that phototropic stem bending towards the light occurs in response to a pigment (probably carotene and/or riboflavin) which absorbs blue light. Some other known blue light effects could involve other pigment systems, or perhaps they represent effects due to weak absorption by phytochrome in the blue region of the spectrum.

Recently, H. Mohr (65), in Freiburg, Germany, has emphasized the importance of a photoreaction which could be unrelated to phytochrome. A number of plant responses are controlled, most of which are also controlled by phytochrome. The pigment system might be different, however, since relatively high intensity illumination is required for relatively long periods of time, and no signs of reversibility have been detected. Furthermore, the action spectra show fairly sharp peaks in the blue (440 mμ) and in the far-red (725 mμ).

Mohr feels that the response under these conditions occurs through the action of a pigment system which circumvents phytochrome control. Phytochrome might also initiate the process in the absence of Mohr's pigment system, and the two could also act synergistically with each other. The Beltsville workers, on the other hand, feel that the responses observed by Mohr are exclusively phytochrome controlled, but that they come about while phytochrome conversion is taking place in both directions under the steady-state conditions which might be expected in high intensity light (*both* forms of phytochrome absorb *some* light at all wavelengths in the visible spectrum — see Fig. 7–3 and discussion on pigweed below).

With the presently available evidence, it may not be possible to distinguish between the two theories. Mohr has recently found an interesting system, however, which supports his theory. Lettuce seedlings produced in the dark have a straight plumule. Irradiation with a small amount of red light causes the cells on one side to grow more than the cells on the other side, and a hook results. Far-red will reverse the effects of red. Thus the hook *forms* in response to phytochrome (the much studied bean hook *opens* in response to phytochrome). After the hook has formed, fairly long irradiation periods with blue or with far-red light at relatively high intensities will straighten the hook by causing the cells *inside* to elongate. It is probably possible to reconcile these observations with a phytochrome theory, but it is quite easy to understand them according to Mohr's concepts.

Whatever the final disposition of theories, the basic observations are most interesting and important. Obviously plants in nature are exposed to long periods of high intensity light, containing amounts of blue, red, and far-red determined by time of day, weather, filtering of forest canopy leaves, and perhaps other factors. Thus study of these matters could be extremely important in gaining understanding of physiological plant ecology. Unfortunately, we have no ideas about how all of this might relate to the flowering process, but it seems certain that important relationships will be discovered (see section on "Other Light Responses" below).

THE KINETICS OF LIGHT ACTION IN PHOTOPERIODISM
(8, 32, 36)

For many years before the pigment was extracted, attempts had been made to understand something about the way that it must act in the flowering process. Using whole plants, effects of light interruptions of different intensities, durations and qualities, given at different times are studied. Since time is usually a variable in such experiments, this is a kinetic approach. The determination of action spectra, described above, is a good example. Much of what is known about phytochrome action in flowering is based upon such studies, which will now be discussed.

1. *The Law of Reciprocity and the Principle of Saturation*

The law states that in a photochemical process it is the total amount of light energy which is effective, as discussed in Chapter 3. There is a reciprocal relationship between time and intensity. Thus a long exposure at low intensity may produce the same effect as a short exposure at high intensity providing the same amount of energy is supplied in both cases. If reciprocity holds, other kinds of experiments become much simpler, so the problem was investigated at an early stage of the research. Nevertheless we still lack complete information on the topic as it relates to flowering.

The method of study is to interrupt the dark period with light of different intensities applied for different durations. A problem arises in how to interpret the data. The Beltsville team approached this problem in the early 1940's by determining the amount of light (time and duration) required to just barely inhibit the flowering of cockleburs or soybeans when light interruption was given in the middle of the dark period. Their results indicated that during the middle 2 hr of a 16-hr dark period, the reciprocity law held. Then in a given experiment (e.g. with the spectrograph) it was possible to vary only the time of illumination at constant light intensity. This has been the experimental procedure used by most researchers for many years now.

One problem of interpretation concerns the saturating amount of light (see Chapter 3) required for the pigment system. As shown in Fig. 7–5, the amount of light required to completely inhibit flowering in the middle of a long dark period (with cocklebur) is less than the amount of light required to saturate the pigment system a few hours before or after the middle of the dark period. It is probable that the amount of light required to saturate the pigment system cannot be determined for the most sensitive part of the dark period. It appears that something less than the saturating amount inhibits flowering completely. The best way to study the reciprocity relationship would be to determine at the appropriate time the saturating light quantity (intensity times duration), by determining the saturating light duration at various intensities. Such an approach is required to accurately evaluate the reciprocity law in flowering.

Surely reciprocity would hold over a range of times and intensities broad enough to validate the early work done at Beltsville, but most studies with biological materials have indicated that when the intensities become extremely high or the times extremely long, the

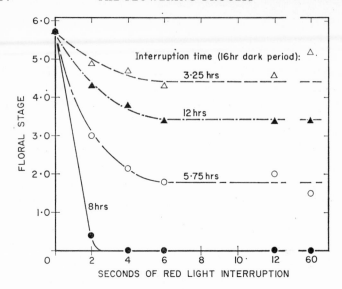

FLORAL STAGE

SECONDS OF RED LIGHT INTERRUPTION

Interruption time (16hr dark period):

3·25 hrs

12 hrs

5·75 hrs

8hrs

FIGURE 7–5

The effects upon flowering of light interruptions given for various durations at various times after the beginning of a 16-hr dark period. Saturation time is about 6 sec, but when the interruption is given at 8 hr a maximum inhibition is obtained with only slightly more than 2 sec. It is unlikely that the 8-hr interruption time measures saturation of the pigment system. Data from Frank Salisbury and J. Bonner, 1956, *Plant Physiol.* 310, 141–147.

reciprocity relationship no longer holds. I have made a few attempts to study the relationship at extremely high intensities and very short durations, using an electronic flash built for photography. Such a light produces extremely high intensities for durations in the neighborhood of 1/5000th of a second (a single flash will inhibit flowering markedly when the light is held a few centimeters away). Preliminary indications were that reciprocity does indeed fail under such conditions.

2. Light Interruptions Applied at Various Times

Many workers using many species have studied the times during the dark period when a light interruption is most effective in inhibition of flowering in short-day plants, or promotion in long-day plants. The experiment is a relatively simple one. A light source of adequate intensity is used, and exposure time is usually held constant. This varies from a few seconds in some experiments to two or three hours in others, and the plant which is used makes quite a difference in the response. Chrysanthemums, for example, are notoriously insensitive to light interruption unless a number of short flashes are given rather than a single long continuous exposure (see Section 5 below). With most plants the experiment is complicated by the fact that a number of inductive cycles are required to bring about flowering, so the experimenter must resign himself to one or two weeks of sleepless nights while he arises at regular intervals to flash lights at his plants (unless he has an automatic set-up such as Hamner's at the University of California in Los Angeles — see Chapter 5). With the cocklebur a single dark period can be used, but if the investigator does many such experiments he may have to shift his metabolism in the direction of night work anyway.

The results of a typical experiment with cockleburs are shown in Fig. 7–6. The light interruptions are most effective near the middle of the dark period. The curve on the left, which indicates increasing

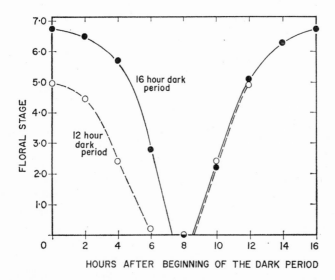

FIGURE 7–6

The effects of a 60-sec red light interruption given at various times during a 16- or a 12-hr dark period upon subsequent flowering. Data from Frank Salisbury and J. Bonner, 1956, *Plant Physiol.* 310, 141–147.

effectiveness as the middle is approached, is nearly a mirror image of the curve on the right, which shows a decreasing effectiveness after the middle has been passed. It seems reasonable to assume that a light interruption near the end of the dark period (the right side of the curve) is like returning the plants to light and is, therefore, very much like studying the effect of different night-lengths (see Chapter 9). A few years ago when the experiment was done, we went on to deduce that an interruption during the first half of the dark period stopped what had gone on up to that time, essentially reversing it so that the processes of the dark period had to start over again. Thus if the plants were interrupted 4 hr after the beginning of a dark period and 12 hr remained following the interruption, it was reasonable to expect that they would flower to about the same degree as plants which either had a 12-hr uninterrupted dark period or were interrupted 12 hr after the beginning. The curves in Fig. 7–6 do seem to fit the interpretation, but there are problems with the interpretation which are discussed in the next section.

3. *Light Interruptions at Various Times Using Cobaltous Ion Treated Plants*

Some time later the same sort of light interruption experiments were done with plants which had been dipped in solutions of cobaltous chloride just before the dark period. Plants treated with cobaltous ion require a longer period of darkness for the first perceptible signs of flowering than untreated control plants (see Chapters 8 and 9). When the light interruption experiment was done, the right hand curve (Fig. 7–7) shifted to the right, as was to be expected from this effect of cobaltous ion on the critical dark period. Reasoning by the above simple theory, it was easy to decide that the left-hand part of the curve should be shifted to the left. A plant interrupted 6 hr after the beginning of a 16-hr dark period should flower about the same as one interrupted 10 hr after the beginning of a 16-hr dark period. As can be seen from Fig. 7–7, however, our expectations were not borne out. Rather than the curve shifting to the left, it stretches towards the right so that it actually crosses the control curve.

To reconcile these facts with the rather simple picture outlined above, one must make some very elaborate assumptions about the action of cobaltous ion, which at this stage of the game do not seem to be justified. An alternative theory comes to mind: the effectiveness

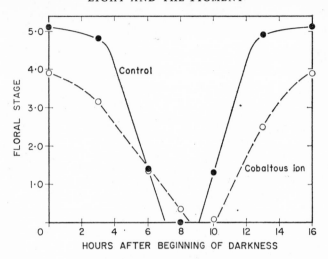

FIGURE 7–7

The effects upon subsequent flowering of a 60-sec red light interruption given at various times during a 16-hr dark period to control plants and to plants previously treated with 3.0×10^{-3} M $CoCl_2$. Dark period initiated on January 16, 1960. Data previously unpublished.

of a light flash must depend not so much upon the length of the dark period preceding or following as upon the time when it is given in relation to the beginning of the dark period. This would seem to imply that some time-measuring mechanism is set into action when the plants are placed in the dark, and that the effect of a light interruption is dependent upon this time-measuring mechanism. If this idea is correct, cobaltous ion must slow down this reaction. Furthermore, a brief interruption of light must not reset the time-measuring mechanism, otherwise we would be back where we were to begin with.

Actually, it is not even necessary to use cobaltous ion to obtain results which are difficult to understand by the simple mechanism proposed in the above section. For example, if a light interruption is given after 8 hr in a 20-hr night, flowering is completely inhibited even though the remaining 12 hr would normally cause maximum flowering if they had not been preceded by the 8 hr and the light flash (Fig. 8–1 in the next chapter illustrates this phenomenon with a 64-hr night). Is this just another demonstration of the requirement for high intensity light preceding an inductive dark period (Chapter 6)?

K

Or does it also show that timing is not reset by a light flash in the middle of the dark period, but that sensitivity to this flash is greatest after 8 hr regardless of how much darkness is to follow? Obviously we have been led into the problem of timing, the subject to be discussed in the next chapter.

4. *Continuous Threshold Light*

It would be of general interest to know what is meant by darkness in the induction of a short-day plant such as cocklebur. How low must the light intensity be before the plant no longer is able to respond to it? The question has bearing on responses to twilight and moonlight; in a reciprocity study such a situation represents the lowest effective light intensity for the longest interval; and response to threshold light has implications relating to our understanding of phytochrome action and time measurement.

The data are relatively simple to obtain. Plants are induced (or inhibited in the case of long-day plants) by placing them at different distances from a point light source in a room with dull black walls. Figure 7–8 shows the results of such an experiment, in which prepared cocklebur plants were arranged on concentric circles around a $7\frac{1}{2}$ watt incandescent bulb and induced with a single 16-hr dark period.

Incidentally, the plants which show essentially no response to the light were far enough from it that they could just barely be seen by the moderately dark adapted eye. The minimum red light intensities to which the plants will respond are probably below the red light intensity of full moonlight (see Fig. 3–2 and 3–4), although it is likely that natural shading by other leaves would allow induction to occur readily even on nights illuminated by a full moon in a cloudless sky. Nevertheless, the degree of response might be influenced, so we do have some precedent for plant response to the phases of the moon (see also 48).

5. *Kinetics of the Far-red Reversal of Red Interruption*

In the early part of the 1950's, R. J. Downs performed a series of experiments at Beltsville, in which he studied the kinetics of the far-red reversal of red inhibition of flowering in cocklebur and Biloxi soybean. One experiment is of particular interest, in which he studied the time interval between red interruption and subsequent far-red

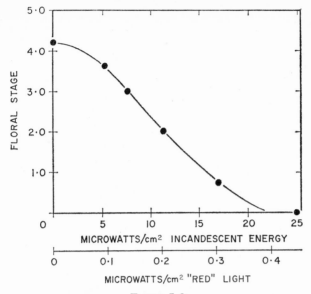

FIGURE 7–8

The effects upon subsequent flowering of low intensity light applied continuously during a 16-hr "dark" period. Experiment initiated on April 26, 1962, by James Whitmore; light intensities measured by Charles Curtis with an Eppley circular 8 junction bismuth–silver thermopile (with and without a Schott interference filter transmitting only in the red region between 641 and 669 millimicrons with a maximum at 655) — both of the Advanced Plant Physiology Class. Data previously unpublished.

reversal. The results with cocklebur are shown in Fig. 7–9. Complete reversal of red induced lettuce seed germination is relatively easy to demonstrate for a number of hours, but in cocklebur and soybean the ability to reverse a red effect with far-red is rapidly lost, so that by the end of 30 min no reversal with far-red can be observed.

More recently, H. A. Borthwick and H. M. Cathey (19), also at Beltsville, have performed related experiments with chrysanthemum, a short-day plant. They found that in order for a continuous light interruption to be effective in the middle of an inductive dark period, it had to be about 4 hr long. If a number of flashes were given, however, the total amount of light required for inhibition proved to be much less. Light flashes given every few minutes for 4 hr were as effective as continuous light given for 4 hr.

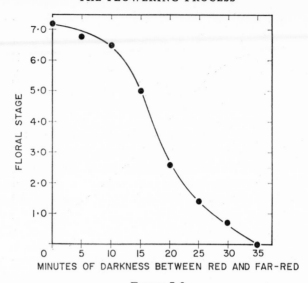

FIGURE 7–9

The effects of far-red light (3 min) given at various times after interruptions with red (2 min) given near the middle of three 12-hr dark periods. Data from R. J. Downs, 1956, *Plant Physiol.* 31, 279–284. The flowering stages of Downs have been converted according to his description to the Floral Stages described in Chapter 5.

Thus it appears that in cocklebur F-phytochrome completes its inhibitory act in 30 min. In chrysanthemum, however, 4 hr is required for complete inhibition, and F-phytochrome decays to R-phytochrome in much less time than this (perhaps on the order of 30 min). To inhibit flowering then, F-phytochrome must be available for about 4 hr, and it can either be maintained continuously by constant light, or it can be renewed at regular intervals by repeated light interruptions.

6. *The Phytochrome System in Japanese Morning Glory* (32)

Although there is ample evidence that phytochrome plays an important part in flowering of Japanese morning glory, it appears from the extensive and detailed work of the Japanese, and recently the Beltsville team, that its specific role is clearly quite different from the role played in the cocklebur. The story becomes quite involved, and the interested reader is referred to the many original papers. Here are some of the basic facts.

Red light inhibits flowering maximally when it is applied about 8 hr after the beginning of the dark period. Far-red not only fails to reverse this inhibition but is itself inhibitory at this time. As a matter of fact, a sufficient intensity of far-red will inhibit as much as will red. Far-red also inhibits, although not as much, when it is applied at the beginning of the dark period, and in some experiments red actually promotes at this time. (In experiments with cocklebur at Beltsville, far-red promoted at the beginning of the dark period.) The inhibitory effects of far-red at the beginning of the dark period can be reversed by red, as can the promotive effects of red by far-red. After about 16 hr (using abnormally long nights) no light interruption has any effect, although in cocklebur highly significant inhibitions can sometimes be observed following such treatment.

At present we cannot choose between various alternative explanations for these results but obviously plant response to light varies considerably from species to species.

7. *Overlapping of the Action Spectra in Pigweed* (19)

There are a number of experimental results which might be very difficult to interpret if we did not understand that the absorption spectra of R-phytochrome and F-phytochrome overlap considerably (Fig. 7–3). Both absorb some light at all wavelengths from 300 to 800 mμ. An excellent and most instructive example is an experiment done at Beltsville using seedlings of *Chenopodium rubrum*, a pigweed. Six days after seeds are placed on moist filter paper in a petri dish, the seedlings will respond to a long dark period, and a few days later the flowering response can be evaluated using a stage system similar to the one used with cocklebur. In the Beltsville experiment, a 16-hr dark period was interrupted in the middle with enough red light to completely inhibit flowering, and then the plants were placed in the spectrum and irradiated for various time intervals. Intensity across the spectrum varied slightly, but not enough to worry about. The results for 4-min and 64-min illumination times in the spectrum are shown in Fig. 7–10. The 4-min curve is a typical action spectrum, closely resembling the far-red spectra of Fig. 7–3. The 64-min curve, however, has been shifted far to the right, and only at 795 mμ is flowering repromoted more than for the 4-min illumination. Thus at long exposure times the repromotion is lost.

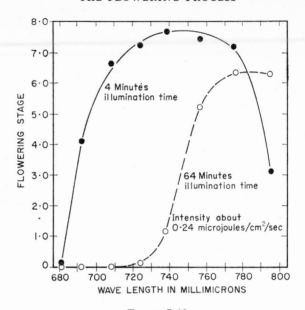

FIGURE 7–10
Action spectra for far-red reversal of red inhibition of flowering in
Chenopodium rubrum, showing effects of either 4 or 64 min of far-red
illumination.   Data   from   Hendricks   and   Borthwick   Australian
paper (19).

If we assume that F-phytochrome is the active, inhibitory form of
the pigment while R-phytochrome has no effect, inspection of the
absorption curves in Fig. 7–3 helps us understand the results shown
in Fig. 7–10. When plants are illuminated at 720 mμ, for example,
most of the pigment is converted to R-phytochrome, but R-phytochrome itself absorbs some light at 720 mμ, and thus some
F-phytochrome is also produced by 720 mμ light. Perhaps 5 to 10%
of the pigment is in the F-phytochrome form even when the plant is
illuminated with far-red light. If the illumination time is short, this
small percentage will be removed by the normal dark conversion in
a short time following illumination. If, however, the illumination
time is long, the small percentage of F-phytochrome will have time
to carry out its inhibitory act, and far-red will be inhibitory rather
than promotive. From Fig. 7–10 it would appear that the percentage
of F-phytochrome remaining is significantly high to wavelengths at

least as long as 775 mμ, becoming more important at shorter wavelengths. Incidentally, the experiment can be interpreted in another way using the High Energy Reaction of Mohr.

8. *Some Other Light Responses* (7, 32)

The literature is replete with studies on the light responses in flowering of both long- and short-day plants. To try to review all of this work here would produce a mass of disjointed narrative which could serve little purpose in a book such as this. The facts are simply not yet all in, and no one has yet been able to produce a comprehensive and coherent picture which will explain in a biochemical and photochemical way all of the reported observations. Nevertheless, some of the phenomena are quite striking, and they certainly further illustrate the breadth of the subject matter. A few examples follow.

Light quality studies, using colored filters and sometimes more elaborate apparatus, have been performed by a number of European workers. Walter Könitz at Tübingen, Germany, found, for example, that far-red light applied during the *light period* inhibited the flowering of pigweed. Of course red light during the dark period also inhibited. The response is a striking one of considerable significance. The far-red effect was reversed by red light — an observation which makes it very difficult to understand how the phenomenon could have been observed in the first place, since plants given far-red light were then returned to white fluorescent light, very rich in red but lacking far-red. G. Meijer on the other hand, found that far-red or blue given during the light period *inhibited* flowering of *Salvia occidentalis*, also a short-day plant. J. A. J. Stolwijk and J. A. D. Zeevaart found the same thing for the long-day plant *Hyoscyamus*. Obviously it is difficult to know what to make of such opposite results. Another example of these perplexing light quality studies is that of Meijer who showed with *Salvia* that the proper mixture of wavelengths during the light period allows red light during the dark period to promote rather than to inhibit, thus in a sense converting it from a short-day to a long-day plant!

Somewhat more clear cut are the studies on whether incandescent or fluorescent light is best for extending the day to promote flowering of long-day plants. With many species, studied by a number of workers, incandescent light seems to be best. As mentioned in Chapter

5 (Fig. 5–6), incandescent light is a mixture of red and far-red (although the red effect predominates in flowering) while fluorescent light contains red but virtually no far-red. Thus it appears that long-day plants, at least, require a mixture of the two forms of phytochrome for optimal flowering. It is clear that extension of the day-length depends upon red light (F-phytochrome). What is the requirement for far-red (R-phytochrome)? It is known that far-red light causes the elongation of stems, and it is also known that stem elongation accompanies the flowering of all the long-day plants used in the above studies. As a matter of fact, it appears that in some species flowering may be caused by anything which will cause the stem to elongate (e.g. gibberellins). Thus it seems that flowering of long-day plants requires F-phytochrome during a long period so that the flower initiating processes may take place, but R-phytochrome (or perhaps *less* F-phytochrome) is also required so that the stems will elongate and the expression of flowering can reach its fullest development.

THE ACTION OF PHYTOCHROME IN FLOWERING

A complete, detailed description of phytochrome's action in the flowering process is still a part of the future, but certain aspects of the problem are beginning to come into focus. We know a little about the form of the pigment which must be active, and we know that it must be acting as an enzyme. But we do not know what its enzymatic activity might be, nor do we know anything about the biochemical steps in flowering which might be influenced.

1. *F-phytochrome as the Active Entity*

The data of Fig. 7–9 seem to clearly indicate that the F-phytochrome produced by a red-light interruption of an inductive dark period (cocklebur or soybean) some way inhibits the flowering process, and that the time required for this inhibition to be complete so that it cannot be reversed by reversing the pigment is about 30 min. In chrysanthemum the time is longer, but the principle is the same. An important difference is that the pigment will decay in the dark to R-phytochrome in less time than is required for the inhibitory act to be complete. In both cases the decay time may be around 30 min, but in cocklebur and soybean this is long enough for complete inhibition, while in chrysanthemum it is not. The results with pigweed (Fig. 7–10) are especially interesting, because they also indicate that F-phytochrome is inhibitory and that this can be expressed by only a small amount of pigment, providing it is available for a long enough time. Various other plants and other systems such as germination also behave as though F-phytochrome were the active form.

Since phytochrome is a protein, it is easy to imagine that F-phytochrome must have enzymatic activity. It might lead to the synthesis or destruction of some essential metabolite. In some systems the quantity of phytochrome required for a given action is extremely small, a characteristic common to many enzymes. For example, a dark-grown pea leaf will show a measurable increase in length (0.1 mm) if only one part in 10,000 of R-phytochrome is converted to F-phytochrome (19). This conversion requires only one erg of incident red radiation per square centimeter. The concentration of F-phytochrome causing this response could be less than 100 molecules per cell.[7]

2. The Role of R-Phytochrome

It is difficult to imagine that R-phytochrome can play anything but a passive role in pigweed. Yet the effect of incandescent light upon long-day plants (stem elongation) could be interpreted by assuming that R-phytochrome leads to the production of a stem elongation factor (gibberellin?). Of course F-phytochrome might inhibit this production, and decreasing it might allow synthesis to proceed. If R-phytochrome does have an effect, then it would appear that optimum flowering in long-day plants requires a balance in the amounts of the 2 pigment forms. The interesting results with Japanese morning glory might also be interpreted in this way, but other, perhaps better, explanations are available. R-phytochrome might play some sort of role in certain systems, but we have no really good evidence that it does.

[7] Assuming a concentration of 10^{-7} moles per liter for total photochrome (the concentration at which it was extracted — see above), the tissue was sensitive to 10^{-11} moles of F-phytochrome per liter. This is about 6×10^{12} molecules per liter. A liter could contain 1.25×10^{11} cuboidal cells, 20 microns on each side (a reasonable sized cell for young tissue). Each of these cells would have 48 molecules at a concentration of 10^{-11} moles per liter.

3. *F-Phytochrome in Flowering*

Two things seem clear: First, the final appearance of phytochrome controlled responses such as flowering must depend on many metabolic steps following the initial absorption of light by phytochrome. This would help us to understand its ability to control so many different physiological processes. It also seems quite likely that many of the final manifestations of its action occur in response to changes in growth regulators such as the auxins or the gibberellins.

Second, phytochrome is a means through which the plant detects the presence or absence of light in the flowering process. Does phytochrome act as an on-off switch, so that when sufficient F-phytochrome has disappeared the flowering timer begins to operate? Or is the dark conversion of phytochrome itself both the switch and the timer? The very success of the threshold experiment (Fig. 7–8) begins to cast some doubt on how phytochrome can act as a switch at all. An on-off switch is an all-or-none piece of machinery, but the threshold response is a quantitative, graded one. There are many fascinating problems in this field, and some of them will be discussed in the next chapter.

CHAPTER 8

TIMING AND THE FLOWERING PROCESS

MEN have probably always estimated the time of day and perhaps have awakened at the same time every morning. Many of us manage to wake up a few moments before the alarm is ready to go off. Yet scientific recognition of this phenomenon of biological time-keeping was rather delayed. Rhythmic motions such as the "sleep" movements of leaves have been noticed by naturalists for the last three centuries, at least. The cocklebur leaf is relatively erect at midnight and nearly horizontal at noon; bean leaves are horizontal during the day and in a drooping position at night. It was easy to "explain" these actions as responses to light or temperature differences between night and day, but some investigators did suspect (in at least one case, as early as the seventeenth century) that these rhythmic movements were a manifestation of an endogenous or internal ability to measure time. The reason for such suspicions was that some rhythmic phenomena will continue even though the environment is relatively constant.

Studies on biological rhythms became increasingly detailed and accurate. Garner and Allard discovered photoperiodism in 1920. Special time-measuring phenomena such as the time memory of bees were discovered (see Chapter 1). Finally in the early 1950's the phenomenon of biological time measurement became recognized by a rather large body of biologists. At this time the idea was so well accepted that workers in the field began to speak of the "biological clock".

The observation which was essential to acceptance of the biological clock idea, was the demonstration that certain time-measuring phenomena (especially the rhythms) would continue even though the organism was held under essentially constant environmental conditions.

MANIFESTATIONS OF BIOLOGICAL TIME MEASUREMENT
(1, 2, 8, 12, 15)

A great many phenomena have been recognized. They seem to fit fairly well into the following four categories: persistent rhythms, photoperiodism, thermoperiodism, and celestial navigation. In a few instances, however, it is quite difficult to classify a given phenomenon. The following outline should give some idea of the breadth of the subject.

1. *Persistent Rhythms*

Probably the most intense interest in biological time measurement is presently centered around these phenomena. Although most of the rhythms approximate 24 hr, only rarely is an exact 24-hr period displayed by an individual organism. More frequently the rhythms deviate up to 1 to 3 hr in either direction from 24 hr. Franz Halberg at the University of Minnesota has suggested the term "circadian" for rhythms which approximate 24 hr in their period. There are many striking examples of circadian rhythms, some of which are listed along with a few comments:

A. Leaf movements in plants. This may have been the first response noticed, and it has now been studied in considerable detail. Much of the study has been carried out with the common bean plant by Erwin Bünning now at the University of Tübingen in Germany.

B. Flower opening and closing, and movement of plant organs other than leaves. Studies of these rhythms have sometimes accompanied work on leaf movements. Some of these phenomena must have been noticed since time immemorial.

C. Color changes in certain crustaceans including the fiddler crab. Relatively recent (1940's) studies on the fiddler crab by Frank Brown and his co-workers at Northwestern University in Chicago represent some of the most intensive and detailed work in this field. These color changes will occur for many weeks in a very dimly lit room at constant temperature, according to superimposed cycles which match the days, the phases of the moon, and the tides at the place where the crabs were collected!

D. Respiration and other metabolic processes. Brown now

studies among other things the changes in respiration rates of potato and carrot tissue, a far cry from fiddler crabs.

E. Growth rates of certain plants. The coleoptile or sheath which covers the first true leaf of a germinating grass seedling has been studied very intensively by plant physiologists because of its growth responses to light and gravity. Among many other things, it seems to grow at rates which vary according to a circadian rhythm.

F. Root pressure. If tops are removed from plants such as tomato or grape, the cut stump will exude sap under some pressure, providing soil moisture conditions and other factors are right. The amount of sap exuded follows a circadian rhythm which seems to be independent of the environment.

G. The activity of certain enzymes, the concentrations of certain pigments, etc. Many such factors in plants and animals vary according to circadian rhythms.

H. Discharge of spores such as those from the fungus *Pilobolus*. Rhythms are easy to study with such organisms, since they can be grown in total darkness, making it easy to control temperature and humidity.

I. Bioluminescence. Certain organisms, especially some found in the ocean, give off light according to a circadian rhythm. The marine green algae, *Gonyaulax*, has been studied in considerable detail. In this case, a population of single-celled organisms is studied rather than one or a few larger organisms. *Gonyaulax* also exhibits rhythms in cell division and photosynthesis.

J. Activity of various animals. Rats, hamsters, cockroaches, and a number of other animals have been studied because of their circadian rhythms of activity. Rats, for example, may be placed in automatic feeding and watering cages with an attached treadmill for running. The treadmill can be monitored through a pen recording on a clock-driven drum so that every time the rat exercises, the time and duration of its activity are automatically recorded. A few dozen of these arrangements can be placed in a room, and the records can be examined after a few weeks. Following a light period, the animals will be active in the darkness (or under low intensity light) about once every 24 hr. Each animal has a very consistent period, which is usually somewhat longer or shorter than 24 hr. Thus after a few days they are all out of phase with each other; some are sleeping while others are running.

K. Insect metamorphosis. A population of fruit flies displays a circadian rhythm in changing from the pupa to the adult stage; some individuals changing at one time and others at intervals of 24 hr thereafter. Colin S. Pittendrigh at Princeton University has made detailed studies of this and many other rhythm phenomena.

L. Mitosis and the size of the nucleus. It has long been known that cells within certain tissues divide predominantly during certain times of day. Constant environment studies indicate that this is sometimes a manifestation of endogenous rhythms. Bünning and his students have also found that the size of the nucleus may vary according to a circadian rhythm.

2. *Detection of the Season by Biological Time Measurement of the Day and/or the Night*

This is the phenomenon of photoperiodism. Phytochrome controlled examples of photoperiodism in plants were listed in Table 7–1 in the last chapter.

A. Initiation and development of reproductive structures. This relates the topic of this book to biological timing in general.

B. Vegetative growth of many plants. A tomato plant grows very poorly under continuous light and constant temperature. Further, if a normal dark period is interrupted for a short time, growth is not as good as in controls given an uninterrupted dark period. This seems to be very analogous to the short-day flowering response, although other observations are not so analogous. For example, if temperature or humidity are varied, good growth may occur in continuous light. Other vegetative responses such as tuber or bulb formation, succulency, etc., are more clear cut.

C. Entering and breaking dormancy. Many plants in temperate regions go into dormancy as the days shorten and the nights lengthen. Coming out of dormancy is also a photoperiodic response in some species.

D. Germination of some seeds. Interestingly enough, some seeds depend not just on the presence or absence of light for germination, but upon the duration of light or darkness.

E. Life cycles of animals. Insects, mites and other animals that undergo stages of metamorphosis from one life-cycle stage to another often do so in response to the length of day. There are many known examples, some of which are quite spectacular. The pitcher plant midge larva floats in the water of a pitcher plant, a carnivorous plant which digests insects in order to obtain a nitrogen supply. This little mosquito-like creature, however, resists all attempts to be digested and goes through its developmental stages in this rather interesting environment. The insect pupates when the day-lengths exceed about 12 hr. The midge responds to light intensities that are at least an order of magnitude dimmer than those to which sensitive, day-length controlled plants will respond. This is near the limits of human vision, so the midges are responding to light when the human observers might agree that it was dark.

F. The reproductive and migratory cycles of many animals. Both invertebrates and vertebrates, including for example insects, fish, lizards, turtles, birds, and mammals become reproductive only at certain times during the year. Often this is a response to day-length, as was first discovered in 1925 by W. Rowan with slate-colored juncos in Canada. Birds were captured during the summer and kept until fall. Then the day-length was gradually increased with artificial lights to simulate the conditions of spring. The birds responded by undergoing the gonadal swelling typical of spring and exhibiting the symptoms of restlessness which are associated with the desire to migrate. When the juncos were released in Canada in the middle of winter, they flapped their wings and headed north into the blizzard, because they had been physiologically "convinced" by the long days that spring was here and that it was time to fly north. Obviously this was an important step forward in science, but it does seem like a rather unethical trick to play on the birds!

G. The color of rodents. Certain animals such as the Arctic hare change color from the brown of summer to the white of winter in response to shortening days. Obviously this is quite closely related to the gonadal development discussed above.

3. *Thermoperiodism*

Greenhouse operators have probably always thought that plants grow best if the natural temperature fluctuations of day and night are simulated. Thus it is common practice to lower the greenhouse temperature at night. Frits Went, while at the California Institute of Technology, demonstrated the reality of this concept, calling it thermoperiodism. He grew various species of plants under many combinations of day and night temperatures, and in a majority of cases growth was considerably accelerated when the night temperature was lower than the day temperature. There are exceptions, including many plants which respond equally well to constant temperatures (such as cocklebur) and even plants which grow best when the night temperatures are somewhat above the day temperatures (such as the African violet[8]). Probably it is not safe to conclude with certainty that thermoperiodism is a demonstration of biological timing, but it does seem likely that this will prove to be true. It may even be a special case of persistent circadian rhythms in sensitivity to temperature.

4. *Celestial Navigation*

This is certainly the most spectacular demonstration of biological timing which has yet been discovered. Certain insects and birds are able to sense direction on the earth's surface by comparing the position of the sun or stars with their endogenous time-measuring system. The honey bee, which has been studied intensively, flies in a given direction to a feeding place by noting the position of the sun, and correcting for the time of day. The pond skater when placed on land always hops south — or the direction that would be south according to the position of an artificial sun set up by the experimenter! Celestial navigation was demonstrated for a European warbler, the lesser white throat (*Sylvia c. curruca*) that migrates at night by placing the birds in a planetarium in Freiburg, Germany, where it was possible to shift the projected "stars" in the "sky", setting up star configurations comparable to various seasons and times during the night. Knowing the directions taken by migrating birds in nature, it was possible to predict the positions they should assume in relation to the planetarium dome, with its projected "stars". The birds responded exactly as predicted. Anyone who has navigated by the stars can appreciate what a fantastic feat this really is. The timing

---

[8] Thus Went generalizes that there are two kinds of people in the United States: those who can grow African violets and those who can't. He further concludes that those who can are those who fail to change the thermostat setting when they go to bed at night, leaving the house warm, while those who can't prefer to sleep in cool houses and not only lower the temperature setting but may throw open the windows before retiring.

must be highly accurate, stars must be recognized, and time of year and time during the night must be compensated for by involved calculations. Nevertheless, the entire system for carrying out this intricate procedure is built into the brain of the bird from birth, since birds raised in captivity are quite capable of responding and migrating as well as birds that have already made the trip.

The ability to train honey bees to feed at a given time of day as well as at a given location may be an example of adding a time memory to celestial navigation. It should perhaps constitute a fifth major category.

Bünning's Theory of Endogenous Rhythms Applied to Flowering (1, 8, 12)

Timing in flowering has been overlooked somewhat in the search for the pigment system, for the flowering hormone, and for other chemical mechanisms. Most workers in the field seemed to feel that timing was merely a matter of how long it took for certain reactions to go to completion. The objective of research was to find the reactions and study them, but nothing out of the ordinary was expected.

This was upset somewhat by Bünning, the German botanist who has studied persistent rhythms. He had accumulated an extensive background of information about the endogenously controlled leaf movements in many plants (especially the common bean), and he had developed the idea that these movements went through a two-phase cycle. One phase could be observed at night when the leaves assumed one position, and this Bünning called the skotophile[9] (or photophobe) phase. The light phase, when the leaves assumed another position, was called the photophile phase. This description and theory of leaf movements was applied to the flowering process. He said that flowering was promoted if plants were exposed to light during the photophile phase. He had shown that the phases were set in most species when the plant went from the darkness into the light in the morning (some species, including the cocklebur, initiate the cycle upon going from light to darkness). It was postulated that long-day plants had a delayed photophile phase, so that a long period

---

[9] Usually in America these terms are spelled without the "e" (e.g. skotophil), but the spelling here is probably preferable.

of light was required following darkness. Thus the plant receives light while it is in the photophile phase and receives darkness only when the cycle had finally shifted to the skotophile phase. Short-day plants, on the other hand, complete their photophile phase in a relatively short time after the lights come on and are beginning the skotophile phase when the night overtakes them. This theory fails to take into account some of the facts of photoperiodism, such as the greater importance of the dark period in relation to the light period, but with a little improvising, most of the facts could be quite well accommodated by Bünning's basic concept.

To me the important aspect of Bünning's theory is that it postulates an endogenous timer which is basically of an oscillating type. The flowering process is supposed to be tied to this endogenous oscillator, which is capable of measuring the length of the light and the dark periods. Of course Bünning's theory is not this simple. The various phases, along with a number of details which we shall not take time to discuss here, complicate it. Nevertheless, the fundamental idea is that the requirement for a specific photoperiod for flowering is controlled by an internal oscillator.

Considerable controversy developed about the theory, which was tested in a number of ways including the following: certain flowering plants were given extended dark periods (2 or 3 days) which were interrupted by flashes of light ("flashes" lasting sometimes from 2 to 4 hr) at various times. According to Bünning's theory, the light should promote flowering if it fell during the photophile phase and inhibit if it fell during the skotophile, and in some experiments this proved to be the case. Furthermore, peaks in promotion (or inhibition) were often separated by approximately 24 hr. Figure 8–1 is an example of such an experiment, using soybean on a 72-hr cycle. On the other hand, other workers performed the same experiment, often with different plants, and were unable to obtain any such clear-cut results. D. J. Carr succeeded beautifully with *Kalanchoe* while G. Hussey failed completely with *Anagallis* and *Arabidopsis*. Bünning said the ones which failed would exhibit persistent rhythms only for 1 or 2 cycles anyway.

As noted above, Bünning's theory stated that the leaf movement phases were set by the onset of the light period. There were some impressive evidences which seemed to relate this to flowering. A series of soybean varieties, for example, exhibited different degrees

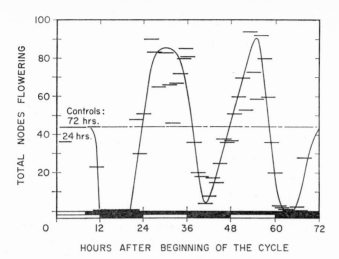

FIGURE 8–1

Flowering response of Biloxi soybean to 4-hr light interruptions applied at various times during a 64-hr dark period. The data are indicated by horizontal lines which show the 4-hr duration of the interruption. Plants received 7 cycles of 8-hr light followed by 64 hr darkness, as indicated by the bar at the bottom. Postulated photophile and skotophile phases are also shown by light or dark bars on the bottom. Data from Hamner (15).

of delay in their leaf movement activities. On the basis of this it was possible to successfully predict whether or not each variety should be a short-day plant or a day-neutral plant. On the other hand, P. F. Wareing in England found with Biloxi soybean and cocklebur that there was no relationship between the effectiveness of a light interruption of an inductive dark period and the time of the previous onset of the light period. Interruptions coming near the end of a dark period might be inhibitory even if they fell in what might be predicted to be the photophile phase. Bünning again pointed out that some plants such as cocklebur were set in their phases by the onset of darkness rather than light, but this failed to explain some of the situations which Wareing and others had discovered. Bünning himself found that some short-day plants and long-day plants exhibited the same photophile or skotophile characteristics at the same time of day. This was probably the most serious retreat which he was forced to make.

In view of all the difficulties which were encountered when one attempted to apply Bünning's theory directly to the flowering process, most workers began to feel that the theory was not a valid one. Unfortunately, the real heart of the matter was largely overlooked: Is timing in the flowering process controlled by an endogenous oscillator or by the time required for various chemical reactions to go to completion?

Is Timing in The Flowering Process Based Upon an Hour Glass Principle?

Cycles occurring under non-varying conditions imply some sort of oscillating timer, like a pendulum or a tuning fork. Bünning says that plants contain such a system, causing them to oscillate with a circadian period between a photophile and a skotophile phase. Flowering in response to day-length is coupled to this system. The alternative idea is that each dark (or light) period is measured by the time required for the completion of a chemical reaction. As the sand flows through the hour glass, so the metabolites flow through the reaction sequence to measure time. Let us consider two strong lines of argument in favor of the reaction rate or hour-glass clock in the flowering process.

1. *The Simplicity of the Hour Glass Approach to an Understanding of Timing*

Plant and animal functions seem to be mostly biochemical. Contrasted with this approach, Bünning's theory of photophile and skotophile phases sounds charged with mysticism. What is the nature of these phases that makes light or darkness influence flowering? In the sense that physiology is the biochemistry of intact organisms, Bünning's theory is hardly physiology at all. Assuming the theory to be true, we are still faced with the problem of discovering the reactions which account for the phases. The flowering process is usually discussed, as it is in this book, in terms of a series of steps. Why shouldn't time measurement simply be the time required to complete the chemical reactions which make up these steps?

There is one serious problem attached to this approach. Biological timing in flowering (as in most circadian rhythms, etc.) is relatively temperature independent, whereas chemical reactions are highly sensitive to temperature. Of course, there are temperature effects in the flowering process, as we discovered in Chapters 2 and 4. The day-length response type itself may be a function of temperature, but the critical dark period is influenced only very slightly compared to what one might expect on the basis of chemical reactions. A typical chemical reaction requiring 8 hr at 25°C might require 16 hr at 15°C, but the critical night for cocklebur changed in one experiment only from 8 hr 48 min at 25°C to 9 hr 4 min at 15°C (see Fig. 9–5 in Chapter 9).

Yet temperature independence is not completely incompatible with the hour glass idea of timing. One simply has to postulate temperature compensating reactions (see below). Although the mechanisms become complicated, they are probably no more so than Bünning's theory. It is interesting to note that circadian leaf movements appear to become adjusted to temperature changes. Time may be measured inaccurately when the plant is first placed in the new temperature (the period is shortened or extended), but after a few days everything is back on schedule, which seems to indicate some sort of compensating system.

2.   The Dark Conversion of Phytochrome as the Timing Mechanism

Sterling Hendricks and other workers at Beltsville (1, 10, 19) have suggested that the time of conversion of F-phytochrome to R-phytochrome in the dark might account for timing in flowering. The half-time for conversion of F-phytochrome to R-phytochrome in many processes is in the neighbourhood of 30 min to 2 hr, as illustrated in Fig. 8–2 for corn seedlings, which shows a decrease in total phytochrome as well as F-phytochrome. With a 2-hr half-life about 8 to 12 hr would be required to decrease the level of F-phytochrome to 1.5 to 6% of its original concentration. If we assume that the critical dark period length is determined by the time it takes for F-phytochrome to decrease to some such level, then we have essentially solved the problem of timing in the flowering process. Perhaps anything above 1.5 to 6% of F-phytochrome is inhibitory to the chemical reactions which occur in flowering (e.g. synthesis of flowering hormone). When the phytochrome has reached this level, then the other processes may be initiated.

The Beltsville workers suggest that some degree of temperature compensation could be achieved in such a system. Rate of dark

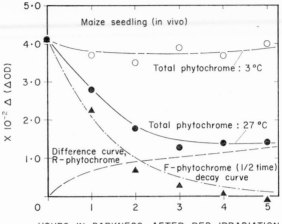

FIGURE 8–2
In vivo measurement of total phytochrome and of F-phytochrome in maize shoots, showing changes with time following a saturating illumination with red light. At 3°C, only total phytochrome is shown. The F-phytochrome curve is a calculated half time decay curve, chosen to approximate the experimental points. Data from Hendricks and Borthwick (19).

conversion of F-phytochrome is decreased at lower temperatures (Fig. 8–2), but then it is less effective at lower temperatures, compensating in some measure for the longer time that it is present. At extreme temperatures such a system would be expected to fail, and time measurement does fail at very low or very high temperatures.

As noted in the last chapter, extracted F-phytochrome is not converted in the test tube in the dark to R-phytochrome. Apparently this conversion requires the presence of the enzymatic systems normally available in the plant cells. Thus the actual decay time for F-phytochrome would be determined by the enzyme systems of the plant under discussion.

An impressive evidence in favor of the idea that timing in general is a matter of F-phytochrome decay times, is the observation that the clock which controls leaf movement seems to be controlled by phytochrome. The leaf movement cycle is initiated when the plants first are illuminated with light following a dark period. Lars Lörcher (61), one of Bünning's students, showed that one hour of red light initiates the cycle in beans, and the effect can be reversed by far-red.

The hour glass theory implies that a red light interruption of the dark period inhibits flowering because it resets the flowering clock.

Figure 7–6 in the last chapter, showing the results of an experiment in which the dark period was interrupted at various times with red light, was first interpreted on the basis that the light interruption resets the clock. Results with cobaltous ion (Fig. 7–7) made this interpretation seem too simple, however. The problem is discussed again below.

Is Timing in The Flowering Process Based Upon an Oscillator Principle?

In spite of the rather simple explanation of timing given above, there is ample reason to believe that the problem is more complex, and that an oscillator might be involved. We are faced with the problem of trying to describe and define how an oscillator might work. The basic idea is that the oscillator is in continuous operation, and the phytochrome system acts in some way to "entrain" or couple it with the flowering process and with the environment. We might imagine, for example, that when the plants are first put in the dark, F-phytochrome begins to change to R-phytochrome, and as the F-phytochrome disappears this begins to remove some block between the oscillating time measurer and the flowering process. Time is then measured by the oscillator until the critical dark period is reached, after which flowering hormone begins to be synthesized (Chapter 9). We can state what must be happening in principle, and we could easily construct a physical model, but it is quite difficult to postulate a satisfactory biochemical explanation. Nevertheless there are a number of observations and ideas which do not fit well into the hour glass theory but can be made to fit the entrainment with an oscillator idea. The difficulties encountered, however, make us wonder if anyone has yet thought of the proper explanation.

1. *The Use of Various Light-dark Cycles or Interruption Sequences*

Karl Hamner (15) at the University of California at Los Angeles and his students have been studying the effects of different combinations of day and night-length upon flowering. They have found that while the soybean, as might be expected for a short-day plant, flowers only when the dark period exceeds a certain length, optimum

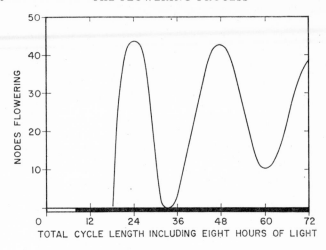

FIGURE 8–3

Flowering response of Biloxi soybean to 7 cycles including 8 hours of light and different dark periods to provide the total cycle lengths as indicated. The curve summarizes six highly uniform experiments (standard errors for the summary points were less than ±0.45 at the 24, 48, 60, and 72-hr points). Data from Hamner (15).

flowering is obtained only when the length of the dark period and the length of the light period add up to approximately 24 hr or multiples thereof. A summary of six of their experiments is shown in Fig. 8–3. Bünning's theory explains the curves nicely, while any sort of reaction sequence mechanism becomes quite complicated in accounting for the 48- and 72-hr peaks in the curve. Interruption experiments, such as that shown in Fig. 8–1, also often imply that the light-dark pattern must match some internal oscillating sensitivity to light or dark.

Könitz' experiment, mentioned in the last chapter, showing sensitivity to red during the dark period and far-red during the light period, also is compatible with an oscillating timer which controls the sensitivity to F-phytochrome (or perhaps R-phytochrome).

2. *Analogy with Persistent Circadian Rhythms*

Circadian rhythms seem to be under the control of an oscillator type clock. Thus the first evidence, the simplicity of the hour glass

approach, loses a great deal of its impact. Rhythms which will maintain themselves for a hundred cycles or more under constant environmental conditions surely seem to indicate the presence of a biological
oscillator. Furthermore, the experiments of Hamner, and other
workers seem to indicate that there is a relationship between this
oscillator and timing in the flowering process.

3. The Sharp Critical Dark Period

It is rather difficult to reconcile the shape of a half-time decay curve
with a sharp critical dark period. As can be seen from Fig. 8–2, a
decay curve is quite flat by the time only 3 or 4% of the original
material is left. A slight variability in decay rate or response to
critical level might make timing very inaccurate. For example, if
the half time for decay is 2 hr, and the critical night is 9 hr, the critical
level of F-phytochrome must be 4.42% of the original. If response
varies within the range 4.42 plus or minus 2%, then critical night
would vary from 7 hr 20 min to 10 hr 45 min. Thus the argument of
simplicity for phytochrome decay timing again loses impact. Such
a system must be so accurate and temperature compensated that it
could hardly be very simple. It is probably just as easy to imagine
that phytochrome decays to a low level very rapidly (within 1 or 2 hr),
and that this then couples the flowering process to a highly temperature insensitive biological timer of the oscillator type.

4. Cobaltous Ion and Clock Resetting (68)

An oscillator clock would allow but does not require resetting by
a flash of light, but an hour glass clock implicitly contains the
assumption that inhibition by a red light flash is effective because
the clock is reset. If the clock is not reset by a light interruption of
the dark period, then the hour-glass type of clock becomes difficult
to accept. In the last chapter, Fig. 7–7 showed results using cobaltous
ion which seemed to indicate that the flowering clock is not reset by
a light interruption. The curve obtained from cobaltous ion treated
plants is shifted to the right and crosses the control curve, so that
after 5 or 6 hr of darkness, plants treated with cobaltous ion are less
sensitive to a light interruption than are control plants. A light
interruption given in the middle of a 16-hr dark period may not
completely inhibit the flowering of cobaltous ion treated plants,
although control plants are always inhibited under these conditions.

When cobaltous ion treated plants are interrupted in the middle of a 16-hr dark period, there are not enough hours either before or after the interruption to allow for completion of the flowering process — yet flowering does occur in most experiments. The effectiveness of the red light interruption seems to depend upon some time-measuring mechanism within the leaf, and this mechanism is slowed by cobaltous ion.

In some of our experiments (but not all) a 6-sec light interruption of the dark period (sufficient to convert all the phytochrome to F-phytochrome) will change the leaf movement pattern. This would normally be considered an excellent example of resetting the clock. Yet in flowering it appears that the 6-sec light flash only inhibits flowering to an extent determined by the timing mechanism, but the clock is not reset. Just how much light is required to reset the clock; that is, to cause all of the dark processes to start over?

Hamner caused soybeans to flower with seven 48-hr cycles (8 hr light, 40 hr dark). The cyles were interrupted at various times with short or with long light periods. Results are shown in Fig. 8–4 (in a similar but more complex experiment low and high intensity light interruptions were used with comparable results). Near the middle of the dark period short light periods failed to inhibit but long periods (or high intensity light) actually promoted far above the control level. Hamner's explanation involving Bünning's theory of phases may apply only to short light periods which fail to reset the clock, but also fail to inhibit during the less sensitive phase of the cycle. Long light periods (or high intensity light) may reset the clock, thus dividing the night into two cycles, so that plants receive 14 instead of just 7 inductive cycles, and thus they are promoted! In preliminary cobaltous ion experiments at Colorado we also have indications that 15 to 20 min of full sunlight may be necessary to reset the clock in cocklebur. Could this be the High Intensity Reaction of Mohr?

What is the biochemical action of the cobaltous ion in slowing the rate of timing? At present we have only one clue. The effect of the cobaltous ion can be reversed by applying it along with various substances known to form complex ions with cobaltous ion. Some of these will completely reverse the cobaltous ion effect on flowering as well as a slight cobaltous effect upon vegetative growth. Ethylene diamine tetra-acetic acid (EDTA) is most effective, causing some

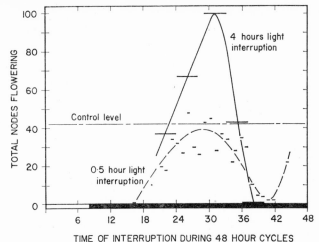

FIGURE 8–4

Flowering response of Biloxi soybean to short- and long-light inter-ruptions given during a 40-hr dark period, during 7 cycles of 8-hr light followed by 40 hr of darkness. Length of the interruptions (0.5 or 4.0 hr) is indicated by the length of the lines which show the data of the figure. Treatments with 1.0 or 2.0 hr (not shown) caused only slightly higher flowering at the peaks of the curves than did treatment with 0.5 hr of light. Data from Hamner (15).

vegetative damage itself, which is reversed by the cobaltous ion! Other compounds such as cysteine also reverse 100%. Aspartic acid reverses about 75%, beta alanine about 10%, and glycine not at all. This may imply that cobaltous ion complexes or otherwise reacts with some substance in the leaf cells which is involved in the timing mechanism, and that the product of this reaction is not as stable as that formed with EDTA, but perhaps somewhat more stable than that formed with beta alanine or glycine. It might now be possible to investigate the cell contents with this information at hand in the hopes of finding some biochemical part of the timing mechanism.

5. *Measurement of the Pigment by Determination of Saturating Light Quantities*

By determining the amount of light required to saturate the pig-ment system, we should be able to measure the quantity of pigment in one form or another. The experiment was discussed in Chapter 7, and the results were shown in Fig. 7–5. Saturation near the middle of the dark period does not measure the amount of pigment (flowering is completely inhibited before pigment saturation is reached), but in various experiments (not all shown) approximately the same quantity of light was required to saturate the pigment system at any time between 2 or 3 hr after the beginning of the dark period and about 7 hr, when complete inhibition is obtained. Thus pigment conversion must be complete within the first 2 or 3 hr, and a timing mechanism based upon pigment conversion will not account for the critical dark period.

In the early stages of work on the reversible pigment system, Harry Borthwick and his co-workers at Beltsville were able to shorten the critical dark period by preceding it with far-red light. But they were only able to shorten it 2 or 3 hr. This is consistent with the idea that the critical dark period is measured only for the first two or three hours by conversion of the phytochrome system, and then some other time-measuring reaction must take over (see also the temperature experiments in Section 7 below and in the next chapter). Even before phytochrome was known, Withrow and his co-workers had shown with other plants that maximum sensitivity to light was reached within two or three hours of darkness, another indication that the maximum amount of R-phytochrome had been obtained by that time.

The indications are that a maximum amount of R-phytochrome has accumulated after a few hours of darkness, but that the sensitivity of the flowering process to F-phytochrome increases continually as the dark period progresses (according to some oscillating timer?). Just before the critical night, only about 30% of the potential amount of F-phytochrome is required to inhibit the flowering process completely (Fig. 7–5).

6.   *The Critical Dark Period under Threshold Light Conditions*

While writing the above paragraphs, it became apparent that an important experiment had to be performed. The critical night should be determined under threshold light conditions. Such conditions imply that the dark conversion to R-phytochrome is not allowed to proceed as far as under conditions of total darkness. An equilibrium is set up in which the tendency to convert to R-phytochrome is balanced by the formation of F-phytochrome by low intensity light.

Using the experimental set-up which was used to obtain the data of Fig. 7–8, the critical dark period was determined under light conditions which cause about 50% inhibition of flowering as measured by the Floral Stage system. Plants were allowed to remain under these conditions for various time intervals, while control

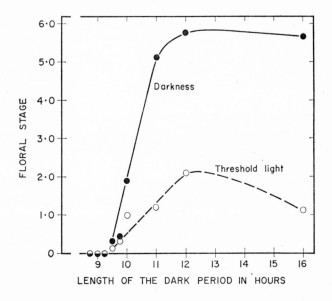

FIGURE 8–5

Flowering response of cocklebur to dark periods of different lengths, given in the presence or in the absence of threshold light (July, 1962). The experiment is typical of four performed with a single dark period and one performed with four dark periods. It is important to note that the actual floral stage of those plants that did flower under threshold light is lower than the Floral Stage of control plants (at night lengths near the critical night, there are vegetative plants in both the controls and the treatments). Thus the inhibition is real and not a statistical artifact produced by more vegetative plants under threshold light. Experiments performed by James Whitmore of the graduate Plant Physiology class. Data previously unpublished.

plants in the same well-ventilated room (uniform temperature) were placed in total darkness behind a black plastic shade. Results typical of four experiments are shown in Fig. 8–5. Although flowering is inhibited under threshold light conditions, the critical night is still the same — timing is not influenced!

The results of this experiment seem incompatible with a phytochrome, hour-glass kind of clock. Timing cannot be a function of the time required for dark conversion of phytochrome if timing is not influenced even when phytochrome conversion is not allowed to go to completion.

On the other hand, the results also seem nearly incompatible with any other available scheme. If the flowering process is coupled with the clock by phytochrome conversion, how can we have any timing at all when phytochrome conversion is not allowed to go to completion? Is it possible that phytochrome has two actions: conversion of only a small amount of pigment couples flowering to the clock (an on-off reaction) but subsequent steps in the process will be quantitatively inhibited (a modulated reaction) unless pigment conversion is complete? It is at least clear from Figs. 7–5 and 7–6 that in flowering sensitivity to F-phytochrome changes during the dark period. Is it possible that this sensitivity has nothing to do with starting or stopping the timing mechanism? Could the High Intensity Reaction of Mohr be the controlling factor here? Such questions must be investigated.

7. *Temperature Effects upon Timing*

It is well known that biological timing is relatively temperature insensitive in most instances. That this is true in flowering is shown by Fig. 9–5 which will be discussed in the next chapter. There is some temperature effect, however. The critical dark period is extended about 20 min by lowering the temperature from 30 to 15°C. Dark conversion of phytochrome, on the other hand, is a metabolically driven, chemical process, quite sensitive to temperature (Fig. 8–2). If the critical dark period consists of dark conversion of phytochrome for up to about 2 hr, after which flowering is then coupled to an endogenous, temperature resistant clock, then one should be able to demonstrate an inhibitory effect of low temperature if it is applied during the first part of the dark period, but not if it is applied a few hours later.

Results obtained previously by various workers seemed to indicate that this was not so. In some cases at least, a 2-hr period of low temperature had virtually no effect upon flowering regardless of when it was given during a 16-hr dark period (Fig. 9–7 in the next chapter is a good example). We decided to approach the problem in a different way. The critical night was determined for plants receiving 2 hr of low temperature (10°C) at the beginning of the dark period, or between the fifth and seventh hours after the beginning of the dark period, or not at all. The results in Fig. 8–6 show that timing was

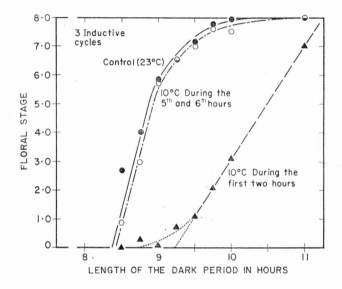

FIGURE 8–6

Flowering response of cocklebur to various night lengths as influenced by 10°C treatments applied during the first 2 hr of the dark period or between the beginning of the fifth and the end of the sixth hours, as compared to controls with no 10°C treatments. Plants were treated with 3 dark periods, beginning July 18, 1962. Temperature during the dark period other than treatment times was 23°C. Data previously unpublished.

indeed slowed by low temperature at the beginning of the dark period, but not by low temperature given later on. It is also easy to see why low temperature inhibition might be difficult to observe with a 16-hr dark period: after 12 hr or so, flowering occurs at a saturation level in spite of the low temperature treatment. These results provide good support for the idea that the phytochrome conversion aspect of timing is limited to a short interval of time just after the beginning of the dark period.

OTHER PROBLEMS RELATING TO TIMING IN THE FLOWERING PROCESS

There are other very important problems in timing besides that relating to the oscillating or hour glass nature of the flowering clock.

1. *Are There One or More Physiological Clocks in the Functioning Organism?*

More specifically, is the clock that controls leaf movements the same as the one that measures the dark period? This may prove to be a rather difficult question to answer. We know that there are a number of superimposed rhythms in living organisms. As mentioned earlier, the fiddler crab has daily rhythms superimposed on tidal rhythms which are superimposed on lunar rhythms. We could build a single clock which exhibited a number of different rhythms. Most clocks in fact do. The second hand has a period of one minute, the minute hand has a period of one hour, the hour hand has a period of 12 hr, and we interpret all of this on the basis of a 24-hr period. Indeed some watches give the day of the year and the phases of the moon. Is this analogous to the physiological timing responses, or does each biological rhythm or manifestation have its own time-measuring mechanism?

2. *Is the Clock Endogenous or Exogenous?*

The discussions so far have implied that timing is controlled from within the organism in a manner analogous to the measurement of time by a wind-up clock. When the rhythms continue, even though the organism is placed in a non-varying environment, the endogenous clock seems very likely. Furthermore, individual organisms vary from one another in the period of their time measurement. Thus the rats running in their cages were out of phase with each other within a few days. How could they be responding to the external environment when each responds with a different period in the same environment?

Nevertheless, Frank Brown (1) feels that time is measured in response to some subtle environmental factor which is not controlled in today's "constant" environment rooms or chambers. He has measured rhythms in rate of respiration of various tissues placed in such rooms, and he finds slight variations in response that are the same for all individuals in the test and that recur at the same time each day. He shows a high degree of statistical significance in his results. Controlling barometric pressure failed to remove these responses. He feels that the organisms are responding to some environmental factor which we are not yet able to control or perhaps even to recognize. Cosmic rays could provide an example. According to this concept, living organisms measure time in much the way that an electric clock measures time by responding to the number of alterations per second in the direction of the electric current. Thus an electric clock may not itself measure time, but the actual time-measuring system is located in the power plant which supplies the electricity. The electric clock only reflects this. Of course, its gear train may be changed so that it runs either fast or slow, and organisms might respond the same way accounting for different periods in the same environment. We shall consider three types of experiments which tend to refute this viewpoint of exogenous timing.

A. Biological time measurement has been shown to continue even when certain organisms are taken into deep salt mines in Germany, where cosmic-ray fluctuations have been essentially eliminated. Of course, there may be some factor other than cosmic rays which was not eliminated.

B. Honey bees have been trained to a time schedule in Europe and then flown by transatlantic jet to New York. If they were responding to some fluctuation in the environment related to the rotation of the earth, then their time measurement should have been upset by this procedure. It wasn't. They continued to measure time on the European schedule even though they were displaced by about 5 hr of earth time.

C. It was reasoned that a way to test the exogenous nature of timing was to remove the organism from the effects of the rotating earth. This could be done, of course, by placing the organism in an artificial earth satellite. Such an experiment has not been done at this time (fall, 1962) but Karl Hamner reasoned that if an organism were placed on the north or south pole on a turntable which rotated once every 24 hr in a direction opposite to the rotation of the earth, the organism would essentially be removed from the effects of the earth's rotation. The creature on the turntable would be moving around the sun (at the earth's speed of 18.5 miles per sec) in a stationary position. If time measurement

M

were dependent upon some exogenous factor resulting from the earth's rotation, then this unfortunate creature should be unable to measure time. Hamner obtained a grant from the National Science Foundation and went to Antarctica during the Antarctic summer of 1960–61. He took with him several organisms which measure time, including both plants and animals. He placed them on turntables which held them steady in space, which rotated in the same direction as the earth, or which did not rotate at all. In every case the organism continued to measure time. His experiment was carried out virtually on the south pole itself, so the rotation of the earth could be nullified. The magnetic pole, however, does not coincide with the rotational pole of the earth, so it might be objected that Hamner's organisms were still in a fluctuating magnetic field, even though other effects of the earth's rotation had been eliminated. At any rate the results failed to support exogenous timing, and Hamner had some extremely interesting months in carrying out his project.

Brown's experiments are valid, whether his interpretation is or not. So how are we to understand his data? Perhaps we can say that the living organism does have an endogenous timing mechanism, but that it is in addition able to respond to changes in very subtle environmental factors. It is possible that we have not yet recognized the factors, and it is highly probable that we have not yet recognized the mechanism of response. Thus is may be possible to superimpose an exogenously controlled periodism upon the normal circadian time measurement of living organisms.

3. Where is the Biological Clock Located?

Single-celled organisms are frequently capable of measuring time, as are the cells of plant and animal tissue cultures derived from multicellular organisms. We may conclude, then, that there are no special organs set aside within the plant or animal for the measurement of time, but that this phenomenon is a function of cellular activity itself.

Where is it located within the cell? Might it be in the nucleus itself? Bünning's students report a change in nucleus volume which seems to follow a circadian rhythm. The period of leaf movement can be changed by treating the plant with chemicals which are known to have an effect upon the nucleus. These include colchicine and urethane (ethyl carbamate). Among many other compounds which have been tested for effects on the period of certain rhythms, the relatively very few which were effective, such as cobaltous ion, ether, various narcotics, etc., may or may not influence the nucleus.

A critical experiment with the green algae *Acetabularia* has recently been performed. This is a single celled organism large enough that the nucleus can be removed, but the functions of the cytoplasm will continue for a while. In such an experiment, time measurement (a rhythm in photosynthesis rate) continued in the cytoplasm in the absence of the nucleus. Thus in *Acetabularia*, at least, timing is not directly dependent upon the nucleus. Can we extend this finding to all organisms?

4. *How Does the Clock Work?*

Many physical models have been suggested to account for and to help explain some of the observational data. Bünning explains some observations by assuming that the biological clock resembles a relaxation oscillator. Low temperature and other treatments may be effective during only a part of the cycle and not during the other part. Bünning suggests that the effect can be observed only when the oscillator is becoming tense.

Colin S. Pittendrigh, and V. G. Bruce at Princeton (1, 18) also propose a rather elaborate physical model designed to account for the experimental observations. Their model, although it is quite complex, is an extremely elegant and impressive one. It accounts for a number of rather detailed observations including some which have not been discussed here. For example, when a rhythm is readjusted by some environmental stimulus such as a light flash or sharp temperature change, the new rhythm may be assumed after several short or several long cycles. Such short or long cycles are called transits.

Yet the model is typically a physical one, and we are still left with no biochemical understanding of the timing process. Discovery of these mechanisms could very well constitute some of the most exciting and interesting research in the future of biology. Probably whole new concepts will arise as we increase our understanding. It is not unreasonable to imagine that the principles learned about timing in living organisms will be comparable in interest and fundamental importance to the recently acquired knowledge about the mechanism of action of the genes.

CHAPTER 9

THE SYNTHESIS OF FLOWERING HORMONE

ONE of the most interesting aspects of the flowering process is the synthesis of a chemical hormonal substance which will result in the control of gene activity and the redirection of growth at the apical tissues. Most of the other aspects of the flowering process (e.g. timing and the pigment system) are quite closely related to other phenomena in biology, and it may well be that synthesis of hormones capable of directing differentiation is also a general phenomenon, but except for a few examples,[10] little is known about such things compared to the other aspects.

EVIDENCES FOR THE FLOWERING HORMONE
(3, 9, 14, 20, 21, 22, 25, 32, 38)

As a matter of fact, we cannot be absolutely certain that a hormone is involved in the flowering process. Such absolute certainty would have to rest upon some understanding of its chemistry, and so far attempts to isolate the elusive substance in a test tube have never succeeded to the satisfaction of a majority of plant physiologists. Nevertheless, three very compelling evidences for such a hormone are presented below, and they allow us to go safely to the brink of absolute certainty in acceptance of the flowering hormone concept.

[10] A similar situation, but one about which much more progress has been made towards understanding, occurs in insects such as the silkworm. Metamorphosis takes place in response to a specific hormone produced by the prothoracic glands. During the very early stages of metamorphosis, a certain region of a large chromosome found in the salivary gland cells enlarges, forming a "puff", while another becomes smaller. It appears that one gene is being turned on while another is being turned off as first steps in the morphogenesis which is about to take place. Starting with 500 kg (more than half a ton!) of fresh silkworms, workers in Tübingen, Germany, isolated 25 mg of pure, crystalline hormone. This material (called ecdysone) not only caused metamorphosis, but the significant changes on the chromosome could be observed within 2 hr after its application! (See U. Clever and P. Karlson, 1960, *Experimental Cell Research* 20: 623–626).

1. The Leaves Respond to the Proper Combinations of Light and Darkness

This is quite easy to show by covering the leaf with a black bag of some sort and leaving the rest of the plant under continuous light. Such an experiment will result in flowering of a short-day plant. Covering the rest of the plant and leaving the leaf exposed will result in the flowering of a long-day plant. It is quite obvious, then, that some response in the leaf is being transmitted to the tip where flowers are formed. I can think of three possible explanations: First, a nervous or electrical impulse of some sort may bring about the transmission. This seems quite unlikely, since such impulses are virtually unknown in plants, and anyway such an impulse is inconsistent with the following two evidences. Second, the nutritional conditions may be upset in such a way that flowering is induced. Perhaps it is a matter of how much sugar gets transmitted from the leaf to the bud. This also seems unlikely in view of the next two evidences, and indeed so far no treatments with sugar, amino acids, or other nutrient compounds has resulted in flowering. Third, a hormone must be involved.

A special committee reported in 1954 with a definition of a hormone which seems quite suitable: Plant (growth) hormones are organic substances formed in one tissue or organ and then translocated to another site where special (growth) control is produced. They must be active in very small amounts, and nutrient elements and energy sources are excluded. If we can eliminate from our thinking a nervous impulse and the upset in nutritional balances, a hormone must be the only alternative left to account for the stimulus sent from the leaf to the bud.

2. The Flowering Stimulus Will Pass a Graft Union From One Plant to Another

Induced cocklebur plants, for example, are approach grafted to vegetative cockleburs as described in Chapter 5. After a week or two the plant which has never received an inductive dark period begins to flower anyway. It is then possible to separate the two plants and graft another vegetative plant onto the one which had been induced by grafting. The third one will then flower. It is reported that this has been carried on through 8 or 10 graft "generations". The long period of time involved surely does not sound like a nervous impulse, and it would be quite amazing if the results could be explained on the basis of nutritional substances. Again we are left with the idea of a hormone, as first suggested by the Russian, Chailakhyan.

A striking aspect of the experiment is that transmission through the graft union will only occur if there is an actual living contact formed between the cells of one plant and those of the other. If the plants are separated by any boundary which prevents living union, the experiment fails even though diffusion of substances through the boundary is possible. The question of why the flowering hormone should move only through living cells has yet to be answered.

Many experiments have been done in which plants of different response types were grafted to each other to see if the products of induction in one type would cause another type to flower. There have been a few failures, but the number of successes is quite astonishing. Thus induced long-day plants will cause vegetative short-day plants to flower on long days, and vice-versa. Day-neutral plants will cause day-length sensitive plants to flower. Present evidence seems to indicate that the end product of induction, the flowering hormone, is at least physiologically the same in virtually all response types.

3. Translocation Rate of Flowering Hormone out of the Leaf may be Measured

Short-day plants which respond to a single night are given a long dark period. Following this treatment one representative group has its leaves removed. A few hours later leaves are removed from another group. This goes on with different groups of treated plants (at perhaps 4-hr intervals) for a few days following the long dark period. When the plants are examined about 9 days after the original dark period, it is found that plants which had their leaves removed immediately following the dark period are quite vegetative. Plants with leaves removed 35 hr or so after the end of the dark period flower nearly as well as plants which never had their leaves removed. Plants defoliated at intermediate times flower to an intermediate degree. This is shown for three experiments with cocklebur plants in Fig. 9–1. It may be possible to imagine other explanations, but surely the logical one is that when leaves were removed immediately following

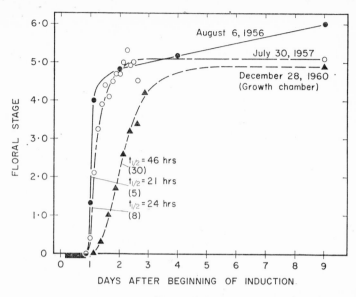

FIGURE 9–1

Three translocation curves obtained by defoliation of plants at various times following a 16-hr inductive dark period. Times of defoliation are shown, and all plants were examined on the ninth day. The lines above the numbers on the abscissa represent noon of the indicated day and the bar on the abscissa indicates the inductive dark period. Approximate times when half the stimulus was out of the leaf are indicated by the $t_{\frac{1}{2}}$. The figures represent half times after beginning of the dark period, and the figures in parentheses represent half times after end of the dark period. Dates refer to the day plants were placed in the dark treatment. Plants represented by the $t_{\frac{1}{2}}=46$ hr curve were kept under 2000 ft-c of fluorescent light in the growth chambers (23°C). Growth chamber data are previously unpublished.

the inductive dark period no hormone (or an ineffective amount) had been transported out. Thirty-five to sixty hours later, all of the hormone had left the leaf and was now in the stem on the way to the bud. This is an excellent experiment to consider in light of the discussion of Floral Stages in Chapter 5, because we must assume that Floral Stage after 9 days is a function of how much hormone reached the tip. In this experiment, as well as in the night-length experiment (see below), the assumption seems reasonable.

FLOWER INHIBITING SUBSTANCES (3, 14, 22, 37, 38)

We can interpret the observation that the leaf responds to the inductive light-dark cycle not only by assuming that a flower promoting substance is made in the leaf and sent to the bud, but by assuming that a flower *inhibiting* (or vegetative promoting) substance is continually sent to the bud under non-inductive conditions, and that the proper environmental stimulus causes a cessation in production of this substance. It is even possible that the *bud* produces such a flower inhibitor which is removed by the leaves under inductive conditions. These interpretations do not fit well with the second two evidences listed above (why should grafting stop production of an inhibitor, or removal of leaves following induction produce results such as those in Fig. 9–1 ?). Nevertheless, there are a number of observations which do support the inhibitor idea for some species. We might summarize in advance by saying that the evidence usually indicates promoting substances, rather often there is also evidence that inhibitors are involved, and in a few cases it appears possible that flower initiation is completely controlled by removal of an inhibitor. The literature on this topic is extensive and somewhat confusing, but fortunately Jan A. D. Zeevaart (37, 38) has recently classified the evidence into three categories as follows:

1. *Interference with Translocation of Flowering Hormone*

In a typical experiment, a single leaf is given the inductive long dark period(s) (e.g. using cocklebur or Biloxi soybean), but a leaf between the induced one and the bud is left under non-inductive conditions. The non-inductive leaf actively inhibits the flowering process. Is this because non-inductive conditions cause the leaf to produce an inhibitor, or do these conditions simply upset the translocation patterns of assimilates within the plant, so that the flowering hormone is unable to get from the induced leaf to the bud? Careful experiments using radioactive tracers have shown that the second explanation is clearly the most likely one. The non-induced leaf does indeed upset the pattern of translocation, and tracers applied to the induced leaf do not reach the bud. This then, is not an example of a true inhibiting substance, but only an inhibiting effect. Yet many reports in the literature base their claim for inhibiting substances upon such experiments.

Incidentally, a number of interesting experiments can be performed to show that the flowering hormone must move with assimilates from the leaf. Thus if one branch of a two-branched plant is induced, the other will flower much sooner if all its leaves are removed so that it is dependent upon the induced branch for its sugar supply. Removing the leaves from the receptor plant of a graft pair also hastens its flowering.

2. *Fractional Induction*

In many short-day plants, long days spaced between the inductive short days cause a positive inhibiting effect. W. W. Schwabe (69) in England has shown in a series of elegant experiments with *Kalanchoe blossfeldiana* that the inhibitory effect of a long day is on the short day which *follows* it. Thus it may require 1.5 to 2 short days to overcome the harmful effects of one *previous* long day. It appears that an inhibitor is produced on long days. The story is more complicated than mere removal of inhibitor by short days, however, since the effects of long days do not accumulate while the effects of short days do. Furthermore, Hamner has shown that the inhibiting effect of long days disappears when the days are shortened, even though the days are still considerably longer than those required to allow flowering. Thus long days must indeed produce an inhibitor, and short days must indeed remove it (and more than one short day may be required for removal), but in such species short days also produce a positive promoting substance which is still being produced long after the long-day inhibitor is gone. The inhibitor probably interferes with production of the flower promoting hormone rather than itself controlling flowering. Zeevaart wonders if the long-day inhibitor might not be F-phytochrome.

3. *Leaf Removal*

If flowering were caused by cessation in production of a flower inhibitor by the leaves, one should be able to cause flowering simply by removing the leaves. With *Hyoscyamus* this critical experiment is successful. Here, however, the story is also complicated by the fact that grafting experiments with *Hyoscyamus* provide evidence for a positive, flower promoting substance. Could this substance be produced by the developing floral bud after flowering is initiated in response to removal of the inhibitor?

After a number of physiological experiments not unrelated to the ones discussed in the above two sections, C. G. Gutteridge in Scotland postulated that flowering in a strawberry variety occurred in response to cessation of production of an inhibitor on long days. Finally, along with P. A. Thompson, Gutteridge (75) tried the critical experiment, successfully causing flowering by removal of the leaves. It even appears possible that the long-day inhibitor in this case might be related to the gibberellins.

In conclusion, it appears that flower inhibiting substances are clearly a reality, if only as a complicating factor. Of course, their importance will depend strongly upon the species, but their presence has been indicated by experiments with short-day, long-day and even day-neutral (pea) plants. The cocklebur is one plant in which there is no clear cut evidence for inhibitors in the flowering process, but in this respect our "type" plant could easily be the exception rather than the rule.

Night Length and Flowering Hormone Synthesis

One of the best ways to demonstrate the kinetics of flowering hormone synthesis in a short-day plant is to treat different groups with dark periods of different durations and then see how they flower after a few days. Cocklebur plants which receive 8 hr of darkness do not flower at all. Plants receiving only one 9-hr period of darkness probably will flower, but the flower buds are very small (slow rate of development) compared to plants which are given still longer dark periods. Such an experiment has been discussed already in relation to timing. Examples were shown in Figs. 3–9, 8–2, and 8–3. Two complete curves obtained with a growth chamber are shown in Fig. 9–2, and curves obtained at various temperatures are shown in Figs. 9–5 and 9–7. The simple way to interpret this experiment is to assume that virtually no flowering hormone is made until the end of the critical dark period, and then the longer plants remain in the dark, the more flowering hormone is synthesized. Under certain conditions, as in the broken line in Fig. 9–2, the first part of the curve has a steep slope for about an hour (part A) and then a sharp break to a much less steep slope (part B). Sometimes the curve is more rounded as in the solid line of Fig. 9–2, but the two parts are still clearly present. This is another instance where we reasonably assume

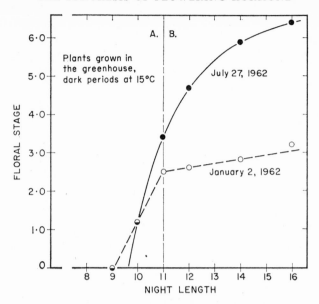

FIGURE 9–2

Two examples of curves showing flowering as a function of night length. Both curves were obtained by placing plants in a growth chamber at 15°C (see Figs. 9–5 and 9–7), but one curve represents flowering of plants grown during the summer in the greenhouse and the other curve shows flowering of plants grown in the winter in the greenhouse. Part A of the curves is the steep part, during which synthesis of flowering hormone appears to be very rapid, and part B shows the period when hormone synthesis is relatively slow. Figures 9–5 and 9–7 show that the amount of hormone may even be decreasing during the B part of the curve at higher temperatures (never, however, during the A period). Data previously unpublished.

that the rate of floral develoment is related to the amount of flowering hormone that arrives at the bud, as discussed in Chapter 5.

LEAF GROWTH AND SENSITIVITY TO INDUCTION

Figure 5–5 in Chapter 5 shows the results of an experiment in which cocklebur leaves of different ages are given an inductive dark period. We concluded that the No. 3 leaf is most capable of synthesizing flowering hormone. Figure 9–3 shows the same flowering data along with the square of the lengths of the various leaves, plotted as

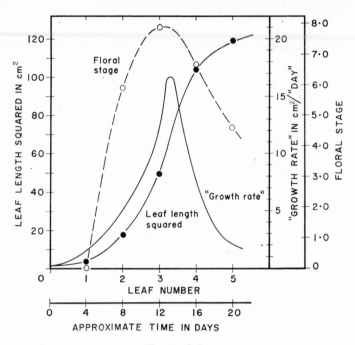

FIGURE 9–3
Relative leaf area (leaf length squared), "growth rate" derived from
this, and Floral Stage as a function of leaf number. See Fig. 5–4.

a function of leaf number. The square of the length is more nearly
proportional to the area of the leaf than is the length itself. If we
imagined that under uniform growing conditions the plants grow
at a fairly constant rate and produce new leaves at a fairly constant
rate, then we might replace the scale on the bottom of the graph with
time instead of leaf number. That is, if a new leaf is produced once
every four days (the approximate rate), then the intervals on the
graph which show leaf number might show time in 4-day periods, as
indicated. The curve then becomes a growth curve, showing size of
the leaf as a function of its age. Such curves have often been obtained
with various organisms.

Plotting the increments of growth during unit time intervals
against age of the leaf, a curve showing growth rate as a function of
time is produced. This curve may also be obtained by measuring the
slope of the size curve (slope is expressed as size units per unit time),

and this has been done in Fig. 9–3. The general shape of this "rate" curve is similar to the flowering curve, leading to the conclusion that the most rapidly expanding leaf is the one most sensitive to photoperiodic induction. The most rapidly growing leaf on a plant is by no means the largest one. It is only a little more than half expanded, but it is expanding more rapidly than any of the others. Perhaps this means that its chemistry is more active, making it more efficient in synthesizing flowering hormone.

Since all of the reasoning in the above paragraph is based on the assumption that the S-shaped curve of Fig. 9–3 is a truly representative leaf expansion curve for cocklebur, it seemed appropriate to check this assumption by actual measurement of cocklebur leaves. In initial attempts during the spring and summer of 1962 we found that the leaves do not seem to grow normally when they are being measured. Figure 9–4 shows the results of our most recent, very careful attempt to obtain a growth curve by daily measurement of midrib length. Although plants were handled only for a few seconds each day, the leaf being measured never grew to the usual 11 or 12 cm length, but turned yellow and died after reaching about 8 cm. Thus we were confronted with the remarkable discovery that one can kill a cocklebur leaf simply by touching it for a few seconds each day! This came as a real surprise, since we have always assumed that cocklebur plants could stand a great deal of abuse with no ill effects. Such results also have some interesting ecological implications. We are presently considering camera techniques in the measurement of cocklebur leaf growth.

Another interesting result of this experiment is evident in the growth rate curve. Although the leaf area curve is approximately S-shaped, there is enough of a deviation from a true S-shape to produce two peaks and a trough in the rate curve. This could be another result of handling, but it might also be the result of changing weather conditions during the period of measurement.[11] If this is true, then we might have some measure of explanation for the difficulties encountered in exactly duplicating experimental results. If sensitivity to a dark period is a function of leaf expansion rate,

[11] Examination of weather data (greenhouse and outside temperatures, daily light conditions) failed to reveal any clear cut correlations with the growth rate curve, although there was a cloudy day with very low light intensity about 4 days before the trough in the growth rate curve.

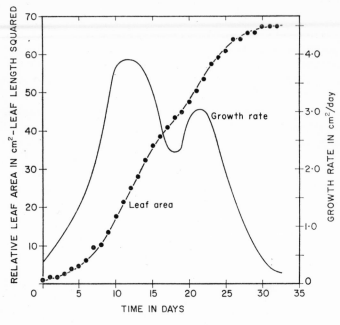

FIGURE 9–4

Size and growth rate of cocklebur leaves as a function of time. Leaves which were 1 cm long at the beginning of the experiment, were measured by Jean Livingston for 33 days, beginning on September 5, 1962. The length of the midrib was determined, but this value is squared in the figure to more closely approximate leaf area. Points represent averages for 10 plants. Growth rate was determined by measurement of the slope of the curve drawn through the points. Data previously unpublished.

then anything which will influence expansion will influence sensitivity. There are, of course, experimental ways to investigate some of these interesting possibilities.

TEMPERATURE EFFECTS

1. *The Inductive Dark Period at Various Constant Temperatures*

Since the very term synthesis implies a series of chemical reactions, we would expect the whole process to be quite sensitive to temperature. Interestingly enough, the subject has very seldom been investigated.

Realizing this, we set up an experiment during January 1962 in our growth chambers in which groups of plants were given dark periods of different durations and at different temperatures from 5 to 35°C. Before and after the dark period all plants were kept at the same temperature. The results are shown in Figs. 9–5 and 9–6. Obviously the flowering process is highly temperature sensitive. We can note, however, that the critical dark period was changed less than 30 min by temperature changes from 15 to 30°C. Even more interesting is the observation that the initial slopes (part A of Fig. 9–2) are essentially the same at the various temperatures from 15 to 30°C. This would seem to indicate that synthesis of flowering hormone during the first extremely active hours of the dark period is not influenced by temperature. No satisfying explanation for this amazing observation is presently available. One wonders if the so-called "synthesis" during those active hours might not be some sort of physical process such as release of previously made hormone from something to which it is bound, or perhaps transport from one location to another. The observation certainly merits further investigation.

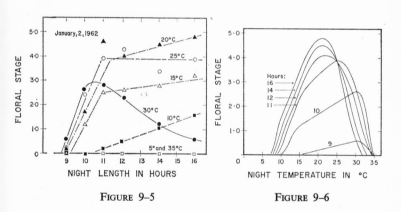

FIGURE 9–5   FIGURE 9–6

FIGURE 9–5

Flowering as a function of night length applied at various temperatures. Temperatures before and after the dark period were all the same (green-house grown plants). Of the 42 measured points, only three do not lie very close to the lines as shown. Data previously unpublished.

FIGURE 9–6

The curves of Fig. 9–5 (without the points) converted so that Floral Stage is shown as a function of temperature for different dark period lengths.

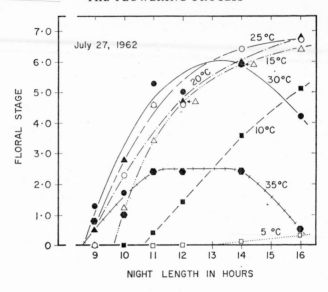

FIGURE 9–7

Data such as those shown in Fig. 9–5 (flowering as a function of night length applied at various temperatures) using plants grown in the greenhouse in the summer. A number of the points fail to fit the curves as shown, and at short night lengths the curves are drawn primarily with reference to those of Fig. 9–5. The trends are very clear, however, at longer night lengths and for lower and higher temperatures. The symbols are the same as in Fig. 9–5. Data previously unpublished.

The curves are different, however, after the end of this initial active period (part B of Fig. 9–2). The low temperature curves (15 and 20°C) are still increasing until the end of a 16-hr dark period, indicating that synthesis of flowering hormone is still continuing during this period. At 25°, however, the curve is level, and at 30°C there is a very sharp decrease in amount of flowering during the B part of the curves.

This decrease can only be understood easily if we assume that flowering hormone, present at the end of the active period, is some way metabolically removed or destroyed or at least made less effective by longer dark periods at the high temperatures. At 10°C this process seems to be essentially ineffective, but the initial synthesis is also reduced. At 15°C the destructive process does not seem to be very important, but the temperature seems to be too low for optimal flowering hormone synthesis. At 20°C flowering hormone synthesis is essentially optimum, and there is virtually no destruction. At 25°C destruction seems to equal synthesis during the B period at 30°C destruction, although delayed for an hour or two during the A period, becomes extremely important, and at 35°C destruction is so predominant (right from the beginning) that no net synthesis can occur.

The experiment was repeated in July, 1962, and the results are shown in Fig. 9–7. The basic observations mentioned above are all evident here, but there are some interesting differences. Unfortunately, effects upon timing are not clear cut. Flowering stage is generally much higher, and the breaks between the A and the B parts of the curves are not sharp. In general, the effects of temperature are less striking, and the temperature range for a given response is much broader. Most of these differences can probably be attributed to the high light intensities and better photosynthesis of the summer months and to the high greenhouse temperatures.

2. *"Interruption" of the Dark Period with High or Low Temperatures*

Temperature has been used as a research tool in much the same way as light to "interrupt" an inductive dark period. For example, B. Schwemmle (1), then at Göttingen in Germany found that flowering was inhibited most in long-day plants when the dark period was "interrupted" near the middle with 3 hr of low temperature (5°C) and least when the "interruption" occurred near the beginning or the end of the dark period. Short-day plants reacted least when "interrupted" near the middle and most when "interrupted" near the ends. High temperatures (35°C) acted in an opposite way.

We have taken this same approach, using our growth chambers, floral stages, etc., to "interrupt" with 2-hr periods of low or high temperature. The results are shown in Fig. 9–8. Our high temperature curve is similar to his, but low temperatures actually seemed to promote flowering, regardless of when applied. It is interesting that high temperatures had virtually no effect until after the critical dark period, after which they inhibited almost as effectively as light. This result has been obtained with a number of species (22). Apparently high temperatures can destroy hormone only after it is made, and timing is resistant to such temperatures.

N

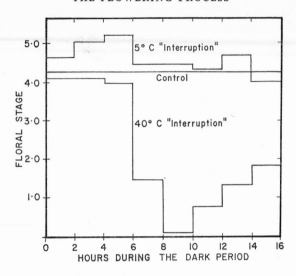

FIGURE 9–8

Flowering as influenced by a 2-hr period of high or low temperature applied during a 16-hr inductive dark period (February 15, 1962). Level parts of the curves represent the 2-hr "interruption" periods. Data previously unpublished.

APPLIED CHEMICALS IN THE STUDY OF FLOWERING (32)

1. *The Problem of Detecting the Flowering Hormone*

Ideally, we should be able to grind up a leaf, keep the enzymes suspended in a buffer solution, and study the synthesis of flowering hormone in the dark. In this way we could learn all about the biochemistry of its synthesis. Such a laboratory approach is presently impossible, because we have no way of measuring the flowering hormone except to observe the flowering of a test plant. Even this would not be a serious limitation if we were able to apply an extract of the flowering hormone to a vegetative plant and cause it to flower. Such a procedure has often been tried. Since about 1938, when the concept of a flowering hormone was first developed, various workers in laboratories all over the world have searched for the proper extract. For example, James Bonner (9) estimates that he personally has made at least 2000 different extracts.

Occasionally we hear reports of success in these experiments. At this time a group at Long Beach State College in California, under the leadership of Richard Lincoln (57), is very excited by a successful extract from cocklebur. The substance E of Nitsch (see Chapter 4) is also interesting. Unfortunately, we now have some rather serious doubts about the meaning of success in this experiment. It may very well be, as with the gibberellins (see Chapter 4), that a substance extracted from plants may lead to flowering when applied to vegetative plants, even though it was not itself a flowering hormone. It may simply trigger the production of the hormone in some way, or the plant may fail to flower because it lacks something *besides* the flowering hormone. Such a substance would be of considerable interest, but it might not fit the rigorous definition for the flowering hormone upon which we must insist: a specific material which is synthesized in one location in response to the proper environment and translocated to the tips, causing flowering.

2. *The Principle of Antimetabolite Action*

In order not to become embroiled in this time-worn problem of extracting the flowering hormone, I tried the more indirect approach of applying chemicals to whole plants and observing the effects on flowering. But what compounds should be tested?

A number of chemicals are known to inhibit certain biochemical reactions. In some cases this inhibition is quite a specific thing. A classic example is the inhibition of respiration by malonic acid, which competes with succinic acid for a position on the enzyme succinic dehydrogenase. Competition occurs because the structures of succinic and malonic acids are very similar, and both may become attached to the enzyme. Malonic acid cannot be dehydrogenated, however, and thus the reaction is blocked.

Let us assume that malonic acid takes no part in any phase of the plant's biochemistry except to block production of fumaric acid by dehydrogenation of succinic acid. Then if we apply malonic acid to a cocklebur plant and flowering is inhibited, we might conclude quite logically that flowering some way depends upon the conversion of succinic to fumaric acid. If on the other hand, malonic acid blocks some other reaction, then our conclusion would not be valid. Furthermore, a plant cell might metabolize the malonic acid so that it never has a chance to interfere with any reaction. In the cocklebur (or the Japanese morning glory) the last possibility seems to be the correct one, since application of malonic acid does not inhibit flowering, even though respiration is known to be part of the flowering process. In spite of such complications, careful use of anti-metabolites might yield valuable information about the biochemistry of flowering.

3. *Reversal of Antimetabolite Inhibition by a Metabolite*

Based upon these principles, we can choose compounds to test in the flowering process (in the manner described in Chapter 5 —

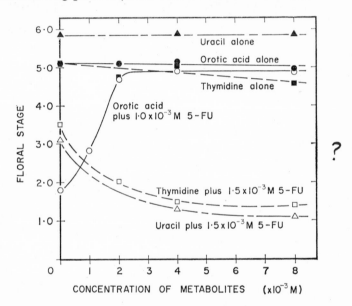

FIGURE 9–9

Effects upon flowering of an anti-metabolite (5-fluorouracil — 5-FU) applied in the presence of increasing concentrations of metabolite (uracil, thymidine, or orotic acid), as compared to effects of the metabolite applied alone. The metabolites alone have no significant effect, and only the orotic acid overcomes the inhibitory effects of 5-FU. Data from Frank Salisbury and James Bonner, 1960, *Plant Physiol.* 35, 173–177.

see Fig. 5–7). We can use anti-metabolites and any other substances known to influence the plant's biochemistry, such as growth regulators or even inorganic compounds such as cobaltous ion. But how can we know what biochemical step is being influenced? With the antimetabolites which inhibit flowering, there is a test. The inhibitor is simply applied in the presence of the suspected metabolite. If as the concentration of metabolite is increased the inhibition decreases, then we can feel fairly certain that the metabolite is a part of the flowering process. Such an experiment is illustrated in Fig. 9–9. There are some complications here, too, but the test is basically a good one.

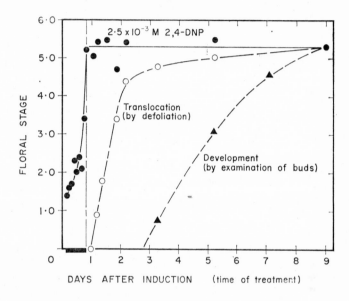

FIGURE 9–10

Effects on flowering of a representative metabolic inhibitor (2,4-dinitrophenol) applied at various times in relation to a single 16-hour inductive dark period. The translocation curve is obtained by defoliation at various times (see Fig. 9–1) and the development curve is obtained by examining buds at various times after the inductive dark period (see Fig. 5–9). Except for the development curve, all points represent Floral Stage of plants examined 9 days after induction (as is the case for virtually all of the cocklebur figures in this book). Data from F. B. Salisbury, 1957, *Plant Physiol.* 32, 600–608.

Actually, there are some technical problems in discovering effective compounds in the first place. Is it possible, for example, that a compound which is applied at the beginning of the dark period might be ineffective, whereas it would be effective if it were applied at some later time in the dark period? Perhaps, but we seldom have time or patience to apply a new compound at any time except the beginning of the dark period. Another problem concerns the possible *promotion* of the flowering process by a compound. If the dark period which

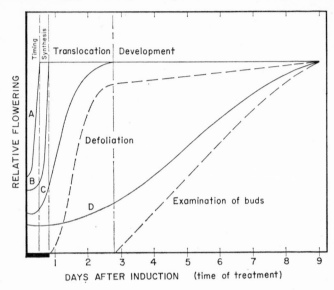

FIGURE 9–11

Summary curves representing many experiments of the type shown in Fig. 9–10. Chemicals which inhibit flowering have been found to be effective when applied during either of the four periods, as shown by curves A, B, C, and D. The A curve represents compounds which are effective only when applied before the end of the critical night (time measurement); the B curve compounds which are effective only when applied before the end of the dark period (synthesis of flowering hormone); the C curve compounds which inhibit if they are applied before translocation of the hormone from the leaf is complete; and the D curve compounds which inhibit floral development regardless of when they are applied. In some experiments, the D curve and the development curve may be nearly the same, indicating that the bud stops its development completely at the time the chemical is applied. Chemicals so far found to be effective in these experiments are summarized in Table 9–1.

we use is so long that maximum flowering (saturation — part B of the curve) is obtained, it may not be possible to observe a promotion. Thus in recent years we have often used dark periods with lengths in the A part of the curve.

TABLE 9–1. COMPOUNDS WHICH INHIBIT FLOWERING OF COCKLEBUR, ARRANGED ACCORDING TO THEIR EFFECTIVE TIMES OF APPLICATION

Compound	Concentration which inhibits 50%	Effect on the critical dark period	Inhibition reversed by metabolite
A. Effective during time measurement (Curve A, Fig. 9–11).			
Cobaltous ion	3.0×10^{-3}M	Extends it	Many compounds known to chelate with Co^{++}(Ch. 8)
SK & F 7997 (tris (2-diethyl-aminoethyl) phosphate) (Zeevaart)	2 mg/ml	none	
5-Fluorouracil (Zeevaart and Bonner)	1.0×10^{-3}	none	Orotic acid
B. Effective during second part of the dark period or before (Curve B, Fig. 9–11).			
1. Respiration uncoupler.			
2,4-Dinitrophenol	2.0×10^{-2}	None	
2. Anti-amino acids			
Ethionine	1.0×10^{-2}	Extends it	Methionine
p. Fluorophenylalanine	2.0×10^{-2}		Phenylalanine
3. Anti-nucleic acids			
5-Fluorouracil (Salisbury and Bonner)	3.0×10^{-3}	None	Orotic acid
Benzimidazole	2.0×10^{-2}	None	Orotic acid and uracil
2,6-diaminopurine sulfate	2.0×10^{-3}	None	Sporadically with orotic acid and uracil, not with adenosine or guanine
8-chloroxanthine	1.0×10^{-2}	None	
4. Miscellaneous Quercetin	1.0×10^{-2}	None	
C. Effective during the translocation period (Curve C, Fig. 9–11)			
1. Auxins			
Indole acetic acid (IAA)	1.0×10^{-3}	None	Slightly by antiauxins
Naphthaleneacetic acid (NAA)	2.0×10^{-4}	None	Slightly by antiauxins

TABLE 9–1. COMPOUNDS WHICH INHIBIT FLOWERING OF COCKLEBUR,
ARRANGED ACCORDING TO THEIR EFFECTIVE TIMES
OF APPLICATION — *continued*

Compound	Concentration which inhibits 50%	Effect on the critical dark period	Inhibition reversed by metabolite
2. Anti-nucleic acids			
4(6)-azauracil	3.0×10^{-3}		Uridine (50% effective or less)
2-Thiouracil	4.0×10^{-3}	None	Uracil and orotic acid
5-bromo-3-isopropyl-6-methyl uracil	7.0×10^{-4}		Orotic acid, uridine, uracil, all partially
3. Miscellaneous			
α-Methyl methionine	2.0×10^{-2}		Methionine
Picolinic acid	1.0×10^{-2}	None	
D. Effective during development (Curve D, Figure 9–11)			
Maleic hydrazide	1.0×10^{-3}	None	
2,2-Dichloropropionic acid (dalapon)	1.0×10^{-2}	None	
2,4-Dichlorophenoxyacetic acid (2,4-D)	1.0×10^{-4}	None	
Thioproline	4.0×10^{-2}		Proline fails to reverse

4. *Time of Application: Determining the Step in the Flowering
Process Influenced by a Chemical*

When we find an effective compound, we must then determine
which step in the flowering process is being influenced. The way to
do this is to apply the compound to various groups of plants at
various times in relation to the inductive dark period. Usually
groups of plants are treated at intervals of two or more hours, beginning at the start of the inductive dark period and continuing until
plants are examined 9 days later (during the last days, plants may
be treated only once each day). The Floral Stage at 9 days is plotted
as a function of the time when the compound was applied, producing
curves such as that in Fig. 9–10. Usually translocation (Fig. 9–1)
and development (Fig. 5–9) are measured in the same experiment.
Figure 9–11 and Table 9–1 summarize experiments with many
compounds.

CHEMICAL SUBSTANCES WHICH INFLUENCE FLOWERING[12]

Effective compounds may be grouped according to the time during the inductive dark period when they are most effective in the inhibition of flowering (we have found a few compounds which promote flowering of cocklebur slightly, but most effective ones are inhibitory).

1. *Compounds Affecting Time Measurement*

We have already discussed in previous chapters the effect of the cobaltous ion on flowering. As can be seen in Fig. 9–11, it inhibits flowering *only* when it is applied during the early hours of the dark period before the critical night. This is the time-measuring phase of the flowering process, and our previous discussions (Chapters 7 and 8) indicated that cobaltous ion inhibits time measurement.

2. *Compounds Affecting Flowering Hormone Synthesis*

Table 9–1 lists a number of compounds which are effective when they are applied during the inductive dark period and which are ineffective when they are applied after the end of the inductive dark period. Most of these compounds become progressively less effective as the end of the critical dark period is approached. This is what we would expect if they were inhibiting flowering hormone synthesis. Compounds that act this way may be further subdivided into at least three groups.

A. Respiration inhibitors. Japanese scientists, using the Japanese morning glory, have found a number of known respiration inhibitors which inhibit during the last part of the dark period. I have shown that 2,4-dinitrophenol acts in the same way. Most of the respiratory reactions continue in the presence of dinitrophenol, but no ATP is formed. Thus flowering hormone synthesis must require the energy of respiration in the form of ATP — which is what one might expect.

In Chapter 6 we reviewed the evidence indicating that energy

---

[12] Effects on flowering of some of the compounds discussed in this section have not been described in previous publications. I am indebted to my colleague, Dr. Cleon Ross, and my former graduate student, Dr. Walter Collins, for much of this work.

substrates such as sugar were essential for flowering. This of course agrees with the conclusions of experiments using respiration inhibitors.

B. Amino acid antimetabolites. Results with some compounds that inhibit during the hormone synthesis part of the dark period are not easy to interpret. For example, a number of compounds were tried which are known to interfere with the synthesis of protein. Most of these amino acid analogs turned out to be ineffective, and we were then tempted to conclude that protein synthesis was not a part of flowering hormone synthesis. One compound (ethionine) was found which might inhibit protein synthesis, although other interpretations were also possible. Some time later another very effective protein inhibitor (p. fluorophenylalanine) proved to be active in our tests. The corresponding amino acids will reverse the effects of these inhibitors. At the time this is being written it would probably be unwise to draw any definite conclusions about the matter, but it surely seems possible that peptide bond synthesis is a part of flowering hormone synthesis after all. Is it possible that the flowering hormone consists of a peptide made up of only a few amino acids?

C. Other compounds. A number of other compounds inhibit flowering only if they are applied during the last half of the inductive dark period or before. In some cases (for example, quercetin) we have little information about how these compounds might be influencing metabolism, but others are known to inhibit nucleic acid metabolism. Since nucleic acids are known to participate in the control of growth and form we were initially very excited upon finding these compounds and discovering that the time of inhibition is the time of flowering hormone synthesis. Again, however, the picture has become quite complicated, and for some years now we have been at a loss to understand all of the implications.

The main complication arises through the observation that 5-fluorouracil (an analog of uracil or thymidine) inhibits flowering when applied to the bud as well as when applied to the leaf. Yet by our definition, hormone synthesis must be taking place in the leaf. Thus if 5-fluorouracil (5-FU) is active *only* in the bud, then its action must be related to transformation of the bud to the reproductive condition (to be discussed in the last chapter) rather than to hormone synthesis in the leaf. The fact that it is effective when applied during the period of hormone synthesis must be some sort of coincidence arising from a delay in its action.

What is the evidence that 5-FU is effective *only* in the bud and not in the leaf as well? Tracer studies seem to indicate that 5-FU will move readily from the leaf to the bud (in small amounts, at least), but not from the bud to the leaf. Orotic acid will reverse the effect of 5-FU on flowering, and its incorporation into ribonucleic acid (RNA) in the bud is inhibited by 5-FU. Thus it seems clear that 5-FU is inhibiting processes in the meristem related to conversion to the reproductive condition.

Is this the only action of 5-FU and similar compounds? We are not yet certain. Sometimes bud applied 5-FU will move to the leaf, at least in trace amounts. Perhaps more important, in preliminary experiments leaf applied 5-FU (which clearly inhibits flowering) has inhibited RNA synthesis in the leaf but not in the bud. When the problems are finally resolved, the anti-nucleic acid compounds may tell us a lot about the biochemistry of flowering.

3.   Hormone Synthesis and the Critical Dark Period

Could time measurement be a matter of the time it takes to synthesize a certain minimal amount of flowering hormone? Probably not, for reasons discussed in Chapter 8 and also because time measurement and hormone synthesis are influenced in different ways by chemicals. Cobaltous ion inhibits during the critical dark period but not later, and furthermore, it extends the critical night. But how can we tell that compounds which are effective during hormone synthesis are not also inhibiting time measurement?

The first approach is to determine the critical night for control plants and for plants treated with the compound in question. Cobaltous ion causes an extension of the critical dark period, as shown in Fig. 9–12. Most other compounds fail to increase the critical night, and we can probably eliminate them as having an influence upon time measurement. Yet two compounds, ethionine and picolinic acid, do extend the critical night. Is it possible that these compounds influence only synthesis of hormone and not timing? If a certain amount of the compound inhibited the synthesis of a certain amount of flowering hormone, then it might take longer to produce the first effective amount of hormone, and critical dark

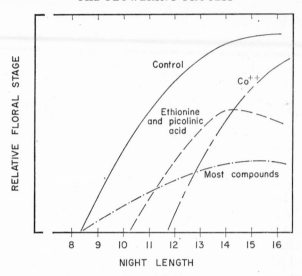

FIGURE 9–12

Summary curves showing the effects on flowering of various substances applied just before dark periods of various lengths. Most compounds fail to change the critical night, even though flowering is inhibited. Ethionine and picolinic acid inhibit flowering at all times by a fairly constant floral stage, thus extending the critical night, and cobaltous ion extends the critical night but does not greatly inhibit flowering at the long night lengths.

period would be extended. How can we distinguish between this effect and an influence upon timing?

The critical experiment which seems to clearly implicate cobaltous ion as an inhibitor of timing is the one in which treated plants are interrupted at various times during the inductive dark period, as shown in Fig. 7–7. The same experiment was repeated using plants treated with picolonic acid, ethionine, or cobaltous ion. Results are shown in Fig. 9–13. Only the cobaltous ion curve is shifted in the manner best explained by an effect upon timing. The other curves are merely lowered, in a manner best explained by a quantitative inhibition of flowering hormone synthesis. Thus effects upon timing and effects upon hormone synthesis can be clearly separated.

4. *Compounds which are Effective during the Translocation Period*

In Fig. 9–11 it can be seen that the auxins cause an inhibition of flowering, even when they are applied after the end of the inductive

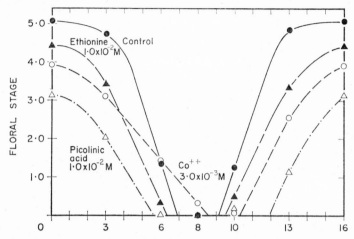

FIGURE 9–13

Effects upon flowering of a one-minute red light interruption (4 photo-flood lamps) applied at various times during a 16-hr inductive dark period to control plants, and to plants treated before the dark period with ethionine, picolinic acid, or cobaltous ion. Data previously unpublished.

dark period, but they must be applied before the flowering hormone has been translocated out of the leaf.

The auxins are hormones which are produced in the stem tip, causing elongation of the stem cells below. Since there is more auxin on the dark side of a stem which is illuminated from one side, and on the bottom of a stem laying down, they offer an explanation for why plants grow towards the light and away from gravity. Since their discovery in 1928 by Frits Went, then at the University of Utrecht in Holland, they have revolutionized research in plant physiology, because, as it turns out, there are few plant processes which are not influenced by them.

In the 1940's it was discovered that applied auxin will inhibit flowering of many plants of all response types. Under certain conditions flowering may even be promoted by auxin. For example, pineapple flowers in response to application of synthetic auxins such as naphthaleneacetic acid. In light of these observations it was theorized that flowering might be a response to changes in the natural auxin concentrations within the leaf. Surely flowering must be influenced by these changes, but at present it seems quite unlikely that the act of induction is mediated through changes in endogenous auxins. Applied auxins will inhibit flowering of cocklebur, provided they are applied to the plant before the hormone is translocated out of the leaf. What is the mechanism of this inhibition?

In explaining the results of the temperature experiment (Fig. 9–5), we postulated a destruction or metabolic breakdown of flowering hormone. Auxin might in some way accelerate this destructive process. Since auxin is known to promote growth, it has been suggested that the acceleration of flowering hormone disappearance is simply a competition between energy requiring growth processes and an energy requiring maintenance of flowering hormone (much like the antagonism theory of flowering discussed in Chapter 4). That is, if energy is continually required to keep the hormone from disintegrating, then anything which would make it more difficult for the leaf to use its energy in this way would result in a destruction of flowering hormone.

It also seems possible that auxin might simply stop (or severely restrict) translocation of hormone from the leaf. The experiment illustrated in Fig. 9–14 seems to indicate that this is not the case. A translocation experiment of the type shown in Fig. 9–1 was performed with control plants and with plants which had been treated with auxin. Although flowering is inhibited, the basic time relationships of flowering hormone translocation are not influenced.

The quantities of auxin which are applied to inhibit flowering of cocklebur also cause vegetative responses. The leaves wrinkle and the petioles twist and curl (epinasty). This in itself seems to indicate (although it does not prove) that the auxin effects on flowering which we observe in such experiments are not typical of the auxin effects within untreated plants.

Other compounds besides auxin act in the same way in the time of application experiments (see Table 9–1). These include a number of inhibitors of nucleic acid metabolism. At present we have no suggestions about how these substances might act in flowering, but the competition for energy idea as applied to auxin may also have some merit here.

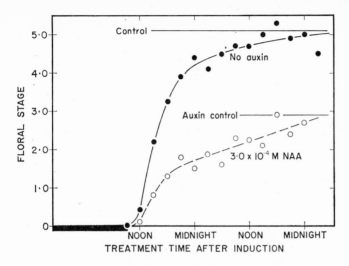

FIGURE 9–14

Effects upon flowering of defoliation at various times following a 16-hr inductive dark period, using control plants and plants treated with auxin. Results are typical of five experiments. Data from F. B. Salisbury, 1959, in R. B. Withrow, *Photoperiodism and Related Phenomena in Plants and Animals*, pp. 381–392, American Association for the Advancement of Science, Washington, D.C.

5. Compounds Which Inhibit Floral Development

Table 9–1 lists a few compounds which inhibit flowering regardless of when they are applied — even if plants are treated just one or two days before the buds are examined. Growth of the flower bud seems to be stopped in its tracks by these substances. The compounds listed are mostly herbicides, so perhaps the response is not surprising. 2,4-D is also considered to be an auxin, but in flowering it acts as an auxin only at very low concentrations.

6. Hormone Synthesis before the end of the Critical Night (32)

Although the biochemistry of time measurement and synthesis of flowering hormone seem to be clearly different, it is not out of the question that the two processes might overlap for an hour or so. Three kinds of experiments may be interpreted as indicating that this is the case. First, applied antiauxins will promote flowering under threshold conditions, and it has been reported that they will slightly shorten the critical night. Is this because a small amount of hormone is normally produced before the end of the critical night, but the natural auxin normally causes it to disappear, while application of an antiauxin preserves it?

Second, high carbon dioxide partial pressures following the dark period will shorten the critical dark period (allow flowering on shorter dark periods). Does this treatment also tend to preserve a small amount of hormone normally produced before the end of the critical dark period?

Third, low temperatures following the dark period will shorten the critical night. These experiments were originally done by workers at the California Institute of Technology (67) and by workers in Holland (78), using a number of inductive dark periods. We repeated them with our growth chambers in Colorado, using either a single dark period or five dark periods. The results, shown in Fig. 9–15, show that the effect can be observed only with more than one cycle. This seems to complicate the interpretation, but it may mean that low temperature following a single dark period tends to inhibit translocation so that flowering is inhibited, even though hormone is not as rapidly destroyed. If a number of cycles are given, preservation of the accumulating hormone may become more important than other effects, and flowering on subcritical dark periods can be observed. Other explanations also come to mind, such as an effect of low temperature on the phytochrome system. Or perhaps we are observing an effect on the long-day inhibitor discussed earlier in this chapter. In Schwabe's experiments, low temperatures made the long days less inhibitory. Perhaps the failure of low temperatures to be effective following a single dark period is because the low temperature only acts on the *following* dark period (by inhibiting synthesis of an inhibitor, such as auxin?).

7. *Cumulative Effects of Repeated Cycles of Induction*

So far all of the discussion in this chapter has been directed towards understanding the processes taking place during *the* inductive dark period, assuming that repeated inductive cycles are only additive but do not interact with each other. It has been suggested that such cycles do indeed interact with each other. R. G. Lincoln and K. A. Raven, two of Hamner's students (58), suggest that the ability of a cocklebur leaf to synthesize flowering hormone increases as the

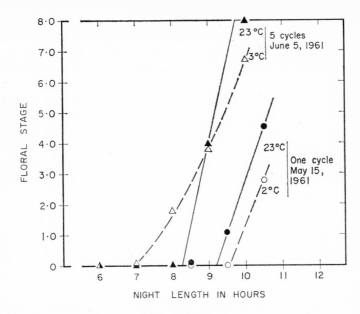

FIGURE 9–15

Effects upon flowering of normal or low temperatures applied for 8 hr immediately following dark periods of different lengths. Effects are markedly different if five inductive cycles are used instead of one. Data previously unpublished.

number of inductive cycles increases. They found that giving one leaf 4 inductive cycles produces a higher Floral Stage than giving two leaves 2 cycles. Of course, the sensitivity of cocklebur leaves changes rather rapidly with time (see Figs. 5–4 and 9–3), and 4 cycles given to the No. 3 leaf would be expected to be much more effective than 2 cycles given to the No. 3 leaf and 2 cycles given to the No. 4 leaf. These workers also found that four long dark periods given over a 10-day period with 2 long days between each inductive cycle were less effective than 4 consecutive cycles. Again, as the leaf ages, it loses sensitivity, and so this would be expected. Proper controls could be used to see if these effects are real, but use of such controls is not mentioned in the published paper.

Schwabe (69), as mentioned above, has investigated the possible participation of an inhibitor in the flowering process. He further interprets his data as evidence for the formation of an adaptive

o

enzyme during successive inductive cycles. Again alternative explanations are available. For example, he shows that once flowering has been induced, nights shorter than those originally required for flowering have a measurable promotive effect on flowering (he used *Kalanchoe blossfeldiana*, but a similar response is indicated in Fig. 3–2). Is this because of adaptive enzyme formation, or do short dark periods produce too little hormone to *cause* flowering, although its effect can be observed when it is added to some already present?

Interactions of cycles or changes in the leaf in response to more than the minimal number of inductive cycles have not been proven, but the likelihood that they exist is great enough to merit our keeping them in mind.

The Synthesis of Flowering Hormone

What does all this tell us about how the hormone is synthesized? Probably our researches have raised more questions than they have answered, but even that may be productive!

There is excellent evidence that a flowering hormone exists, although the picture is often complicated by the presence of inhibitors as well as promoters. Hormone synthesis begins near the end of time measurement, most effectively in the most rapidly expanding leaves. The initial rate of hormone synthesis is not strongly influenced by temperature; an unexpected finding that may lead to new ideas about hormone synthesis. Later phases of the process are highly temperature sensitive, and a destruction of hormone at high temperatures is evident.

Applied chemicals, especially antimetabolites, may help us learn about the biochemistry of hormone synthesis. Time of application experiments are essential to pinpoint the part of the process which is being influenced, and since these are only possible with a plant which will respond to a single inductive cycle, it is easy to see why cocklebur and increasingly Japanese morning glory are emphasized so strongly in this discussion.

Such experiments have shown that respiration (ATP production) is essential for hormone synthesis, and that amino acid (peptide formation) and nucleic acid metabolism may be involved. There is much to learn here, however, and we especially need clarification about the specific roles of the leaf and the bud. This is the frontier.

With some effective compounds, we don't even suspect what is going on, but we do have experiments to investigate whether timing or hormone synthesis is being affected.

Auxins do not seem to play a controlling role in the process, but they do have an effect. A number of other compounds simply stop development of the floral bud, and perhaps they could be profitably used in attacking the problems of the final chapter.

CHAPTER 10

THE MOVEMENT AND ACTION OF THE FLOWERING HORMONE

SOME of the topics which relate to processes taking place subsequent to synthesis of flowering hormone have already been alluded to in previous chapters. Thus translocation of the flowering hormone and its activity in causing the bud to become reproductive have been mentioned a number of times. An interesting part of the flowering process which has not been discussed in any detail, however, concerns the status and activity of the flowering hormone after it is synthesized during the dark period. We will discuss five aspects of this topic.

TRANSLOCATION RATE OF THE FLOWERING HORMONE
(3, 32, 38)

As shown in Fig. 9–1, measurement of the rate of flowering hormone movement out of the leaf can be made by defoliating different groups of plants at various time intervals following an inductive dark period. Experiments such as this have indicated that movement out of the leaf may begin within 2 to 4 hr after plants are moved from darkness to sunlight; that this period of translocation often requires anywhere from 12 hr to 2 to 3 days; and that the rate of movement is probably rather slow. Figure 10–1 shows that the rate is highly temperature dependent, as would be expected for movement in the phloem.

A defoliation experiment is, not however, a good way to measure the linear *rate* of flowering hormone movement, since we are not making the measurement over a linear distance. We are only determining a portion of the hormone which has entered the stem at any given time interval, providing there is sufficient to cause flower development. Other slightly more direct methods have been used to measure rate of movement of the hormone over a linear distance. One way is to utilize two-branched plants, in which one

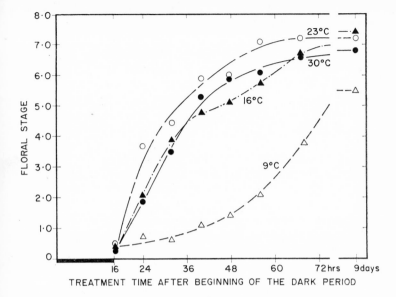

FIGURE 10–1

Effects on flowering of cocklebur of defoliation at various times and
temperatures following an inductive dark period. After 16 hours in the
dark at 28 to 29.5°C (beginning September 27, 1962), plants were transferred to growth chambers at the temperatures shown for 52 hr, after
which they were returned to the greenhouse. Plants were defoliated
while in the growth chambers at the times shown. Approximate half
times for translocation of the flowering stimulus from the leaves are:
9°C = 62 hr after beginning of the dark period; 16° = 32 hr; 23° = 26 hr;
and 30° = 32 hr. As temperature increases towards 23°C, Q_{10} for translocation decreases from about 2.5 to about 1.5. Data previously unpublished. For similar results with Japanese morning glory, see
Zeevaart (37).

branch is induced to flower and the time required for the other branch
to flower is measured. Such methods are also at best only estimates.
They seem to agree, however, that the rate of movement of flowering
hormone is exceptionally slow, in the neighborhood of 2 to 4 mm
per hour. Sugars produced by photosynthesis in the leaf are known
to move through phloem tissue at much faster rates in many plants
including cocklebur, usually between 200 and 1000 mm per hour.

A number of experiments have indicated that flowering hormone
moves in the same direction as the sugar, and movement of the
hormone may be dependent upon movement of the sugar (see

Chapter 9 and below). Flowering hormone moves through the plant only through living tissues, and it will not pass any sort of diffusion boundary (see Chapter 9). We do not quite know what to make of these observations, but it has been suggested that the flowering hormone might be a very large molecule, such as a protein or even a nucleic acid. This could conceivably account for its very slow rate of movement and perhaps for some of the other phenomena which are discussed below.

HIGH INTENSITY LIGHT FOLLOWING THE INDUCTIVE DARK PERIOD (22, 32)

For many years it has been known that cockleburs flower best when the inductive dark period is followed by high intensity light. An experiment shown in Fig. 10–2 illustrates the phenomenon. At least two possible explanations for this finding come to mind.

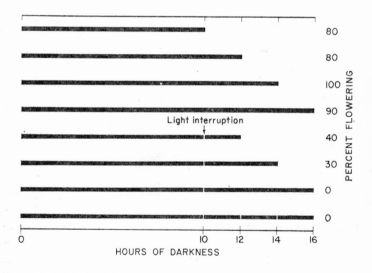

FIGURE 10–2

Effects on flowering of different dark period lengths (apparently under sub-optimal conditions) or different dark period treatments following a 10-hr inductive dark period and a short light interruption (2 min of fluorescent light). Data from James A. Lockhart and Karl C. Hamner, 1954, *Botan. Gaz.* 116, 133–142.

1. *High Intensity Light may be Required Directly to Stabilize the Flowering Hormone or Complete its Synthesis in Some Manner*

Karl Hamner and his graduate student, James Lockhart (59), felt that this was the most likely explanation. They performed a number of interesting experiments in which they studied the effects of temperature, low light intensities, and applied auxins on the processes following the dark period.

2. *Flowering Hormone may be Destroyed in Certain Leaves, and High Intensity Light may Overcome this Potential for Destruction*

In Chapter 9 we interpreted high temperature and applied auxin effects by postulating a destruction of flowering hormone in the leaf. Does high intensity light remove this potential for destruction (e.g. by lowering the auxin level)? Or does high intensity light cause a movement of flowering hormone out of the leaf before it can be destroyed? It is interesting to note that experiments such as that shown in Fig. 10–2 do not always succeed; that is, the inhibitory effect of a second dark period cannot always be demonstrated. Thus it seems that environmental factors may some way condition the response. Is this an effect upon the leaf's ability to destroy hormone?

D. J. Carr (46), then at the University of Melbourne in Australia, was able to overcome the requirement for high intensity light following an inductive dark period by treating plants with sugar. This seems to indicate that the requirement for high intensity light following the dark period is a requirement for sugar produced by photosynthesis. But what is the sugar doing? It might be acting as an energy source for metabolic stabilization of a hormone precursor (according to the idea of Lockhart and Hamner), or for metabolic reactions which remove the destructive potential of the leaf (an idea which I favor), or it might be providing a translocation stream for movement of the stimulus out of the leaf (as Carr suggested). The translocation theory also implies a destruction, however, since plants are finally returned to high intensity light anyway, and if there were no destruction during the second dark period, flowering would at most be slightly delayed.

One point seems clear and is agreed upon by all investigators: the requirement for high intensity light following the dark period is not absolute. As mentioned above, it is often difficult to demonstrate it, and it is also well known that some cocklebur plants will flower even if they are left in darkness continually until the buds are examined one or two weeks later.[13] Furthermore, an effect of high intensity light following an inductive dark period cannot be demonstrated at all in the Japanese morning glory, nor in Perilla. Thus the so-called Second High Intensity Light Process in cocklebur is not typical of plants in general. It seems to be a situation which sometimes complicates the events which lead to flowering of cocklebur. Recently, Norman E. Searle (70) of E. I. Du Pont de Nemours and Company has performed a number of experiments on translocation of the stimulus out of the leaf. Cockleburs were grown under completely artificial conditions, watered with nutrient solutions, etc. Plants that grew poorly under continuous light and constant temperatures were discarded. Induction was performed by darkening a single leaf of plants with the tip removed (the axillary bud grew out to provide the measure of flowering), while two other leaves left below the darkened one remained in continuous light. Under these conditions Searle found no inhibitory effect of darkness or low intensity light following induction. Furthermore, translocation rates of the hormone out of the leaf were essentially the same under high intensity light (2000 ft-c fluorescent plus 3.5% incandescent), low intensity light (0.3 ft-c through small holes in the box used to cover the leaf), and darkness.

These experiments seem to support the findings of Carr. The two leaves left in the light might be expected to supply the darkened leaf with sugars. Thus Searle's experiments seem to be a rather complicated but very efficient way of arriving at the situation which Carr tried to provide: a leaf kept in the dark is supplied with a continuous ample amount of sugar, and the requirement of that leaf for high intensity light following induction is thus obviated.

The effects upon translocation are especially interesting. It has clearly been shown that a darkened (or non-photosynthesizing) leaf acts as a sink for sugars produced by photosynthesizing leaves. Thus sugar must be flowing into Searle's darkened leaf. Yet most experiments up to now have shown that flowering hormone moves only

[13] Of course, there is always a limit to how long plants can be left in the dark; eventually as all their nutrients become depleted they die. Actually, many species will flower in continuous darkness. These include short-day plants, long-day plants, and day-neutral plants. The topic is itself an interesting one, although it will not be discussed here (see 32, 37, 38).

along with the sugar. Searle's experiment may mean that it is not so much a matter of moving *with* the sugar as it is having an ample supply of sugar to activate the movement. Tracer studies should be done with Searle's arrangement to clarify this matter.

Unfortunately, Searle concludes in his discussion that the status of the hormone as well as its translocation is independent of light conditions following the dark period, even though the majority of his experimental plant was left in the light all the time! Zeevaart (38), in discussing these results in a recent article, implies that there never really is an effect of light conditions following the inductive dark period, even though a number of papers by Lockhart and Hamner and various other workers clearly demonstrate such effects. Zeevaart further implies that since Searle's conditions were controlled, his results must take precedence over all others. Can we really put that much faith in controlled conditions?

It should be quite evident from Figs. 9–1 and 10–1 that there is considerable variation in times of translocation from the leaf. This is undoubtedly due to environmental influences (the genetics of the plants were the same in all experiments), and studies with controlled environments should be directed towards understanding the nature of these influences. Just because no variability shows up when experiments are done under one single set of conditions, we should not conclude that results under those conditions are typical of flowering in general.[14] The typical controlled environment is an extremely artificial situation anyway. Light quality and intensity differ markedly from natural conditions. In studies on different photoperiod lengths, the continuous high light intensities provide different photosynthesis times as well as different durations of light. In short, we must take care in using this valuable new tool.

At any rate we might conclude that if a leaf[15] is stopped from making hormone by a light interruption (producing F-phytochrome) and is then left in the dark or low intensity light until its carbohydrate supply is diminished (and not supplied artificially with sugars), then

[14] For example, under a given set of temperatures and light intensities, we can repeatedly observe about a 2-fold promotion in Floral Stage in plants treated with the inhibitor 5-FU. Since most experiments with 5-FU show an inhibition, we would be rather unwise to conclude that 5-FU is a general flowering promoter. Our problem is to try to understand these unexpected results.

[15] The effect of light intensity following the dark period is clearly on the leaf and not the bud.

the amount of flowering hormone reaching the bud is decreased. This might be due to a requirement for sugar to stabilize the hormone, prevent its destruction, or move it out of the leaf before it is destroyed.

INDUCTION AND THE NATURE OF FLOWERING HORMONE
(3, 14, 20, 25, 32, 38)

The flowering hormone is exceptionally interesting in one respect: once it is formed in certain plants it seems to catalyze its own subsequent, further synthesis. This is beautifully illustrated by the cocklebur, which changes from the vegetative to the reproductive condition after a single inductive dark period. Furthermore, if a young cocklebur plant is given one or a series of long dark periods so that the plant becomes reproductive, the young leaves which grow out in the weeks or even months to come are themselves capable of causing flowering. These young leaves, which did not exist at the time the plant initially received its long dark period (or were too small to respond), can be grafted on to a vegetative plant which has never been exposed to an inductive dark period, and they will cause that plant to flower. Thus it appears that the flowering hormone is not diluted out as the plant grows but seems to increase or grow along with the plant. This is demonstrated in another way by passing the flowering hormone through a number of graft "generations", as mentioned in Chapter 9. We get the impression that flowering hormone, once it has been synthesized within the plant, is continually synthesized in all of the new cells that appear through the subsequent growth processes.

This is the sort of thing that the chemist calls autocatalysis, a not uncommon chemical phenomenon. But in the case of autocatalysis in which the presence of a molecule causes the further conversion of precursor molecules into molecules like itself, it can be seen that the end result will always be a maximum concentration of the auto-catalytic molecule (see Fig. 10–3). Thus in autocatalysis the final quantity of the product is not a function of the initial amount of this material present but only of the precursor available for its synthesis. Of course, the *time* required to reach the maximum amount will depend upon how much of the original molecule was present, but the final amount will not.

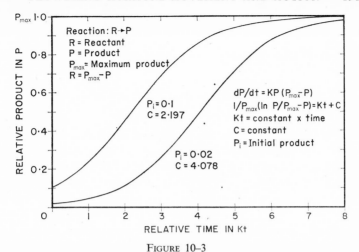

FIGURE 10–3

Two curves showing the amount of product as a function of time for an autocatalytic reaction in which rate of the reaction is a function of the amount of product and the amount of precursor present at any time. The two curves are for different amounts of initial product.

This is not analogous to induction of cocklebur. The initial amount of flowering hormone seems to be determined by the length of the inductive dark period (and, of course, other factors which may influence its destruction, etc.), and under good growing conditions this amount is maintained throughout the life of the plant. If a plant is given a maximum amount of induction resulting in a maximum amount of flowering hormone, development of flowers and seeds will be very rapid (in the case of a cocklebur the development may be complete in about 30 days under continuous short-day cycles). But if a plant is given only a minimum induction (a single 9- or 10-hr night in the case of a cocklebur), development of the flowers will be extremely slow and may require many months for completion (Fig. 10–4). Furthermore, the number of seeds on the plant will be far fewer in the second case than in the first. Thus we have something different from the autocatalytic mechanisms of the chemist. Flowering in cocklebur is an example of homeostatis, in which an initial concentration of hormone, following its establishment by induction, causes a given rate of development which is maintained throughout the subsequent life of the plant — even though the plant continues to increase its volume by growth.

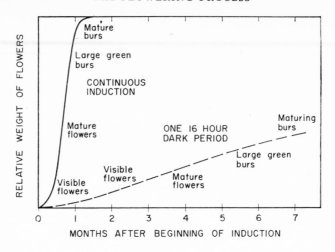

FIGURE 10–4

A schematic diagram illustrating the development of cocklebur flowers and fruits under conditions of continuous induction or following the minimal induction of a single dark period. Up to 3 months the curves are based on the observations of Francis L. Naylor (*Botan. Gaz.*, 1941, 103, 146–154). Following this time the minimal induction curve is based on the familiar observation that such plants will produce ripe seeds, although only after an extended period, and that the size and quality of seeds is much lower than in the case of maximally induced plants. It is interesting to note that in Naylor's experiment, minimally induced plants produced about 10 staminate and 6 carpellate flowers per plant, while maximally induced plants produced about 4 staminate and 12 to 13 carpellate flowers per plant.

I wondered at one time whether the rate of development of a flower bud depended upon the amount of flowering hormone available for that bud or whether it depended upon the concentration of flowering hormone in any given plant tissue. To find out, I prepared a number of plants so that all of them had a single No. 3 leaf. On half of the plants, all of the buds but one were removed, while on the remaining plants all of the buds were left intact. The entire group was then given an inductive dark period, and 9 days later all buds were examined for their flowering condition. Table 10–1 shows the results. The average Floral Stage was not greatly different on the two sets of plants, but of course, the number of flowering buds was much greater on the plants with the most available buds. This seems to imply that Floral Stage or rate of floral bud development

TABLE 10–1. FLORAL STAGE AS INFLUENCED BY THE NUMBER OF BUDS ON EACH PLANT[1]

	Sixteen plants per treatment prepared 3 days before a single 16-hr dark period (June 24, 1953) by removing all leaves but the No. 3 leaf, and removing the stem tip above the No. 3 leaf; buds left on the plant as follows:	
	Only the bud axillary to the No. 3 leaf	All buds
Number of buds examined 9 days after induction	64	219
Floral Stage of terminal inflorescence on shoot axillary to the No. 3 leaf	6.8	6.0
Floral Stage of all buds	6.1	4.4
Product of Floral Stage of all buds times number of buds examined	392	958

[1] Data previously unpublished.

is mostly a function of concentration of flowering hormone within the tissue. If this were not the case, the plants which had the large number of buds should have had a much lower average Floral Stage, since each bud would have had only a much smaller share of the available hormone.

Perilla is also capable of the induced state (although a number of dark periods are required), and induction is quantitative and homeostatic (37, 38). Yet in Perilla the induced state is extremely localized. An induced leaf has been grafted to seven consecutive receptors, causing each to flower, but leaves of the receptors when grafted to other receptors, never caused them to flower. In Perilla the ability to produce flowering stimulus is so localized that one part of a leaf may be converted to the induced state, while the remainder of the leaf remains non-induced. Thus if Perilla plants are induced with a series of short days and then returned to long days, they will flower so long as the induced leaves remain on the plant. When these finally grow old and die, the plant again becomes vegetative. This appears to be one neat mechanism for attaining the perennial condition.

THE REQUIREMENT FOR ACTIVE BUDS (32, 38)

There has been much speculation about what happens within the plant when it is converted to the induced state. A cocklebur plant seems to be incapable of completing the act of induction unless active bud tissue is present. If all of the buds are removed from a plant, which is then given a series of long dark periods, the plant is incapable of causing a vegetative receptor to become reproductive when it is grafted to it. Even if the active buds are removed just prior to induction, and dormant buds are left on the plant, they will usually be vegetative when they finally become active (Fig. 10–5). In order to be reproductive they must become active within 5 or 6 days after

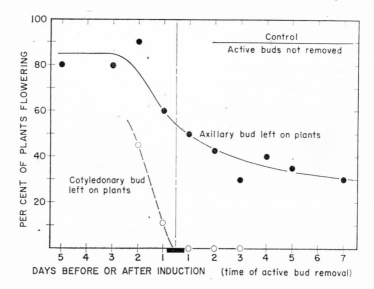

FIGURE 10–5

Effects on per cent flowering of removing active buds at various times before or after an inductive dark period. One curve was obtained by using plants with all buds removed at the times shown but the cotyledonary ones, which were then allowed to become active so that they could be checked for flowering. The other curve shows results obtained by removing all buds but a single relatively dormant axillary one. Note that removal of buds which are already flowering (beginning about 3 days after induction — Fig. 5–9) still causes the dormant ones to be vegetative when they become active. Data from F. B. Salisbury, 1955, *Plant Physiol.* 30, 327–334.

the leaves have received the inductive dark period. Apparently the hormone disappears if it doesn't find an active bud within that time, and leaves lose their ability to produce it. As a matter of fact, plants with developing flowers will revert to the vegetative condition if all the active buds are removed.

This is not true in the case of Perilla. Leaves detached from the plant can be induced and kept for many weeks until they finally die without losing their ability to cause a vegetative plant to become reproductive when they are grafted onto it. But then Perilla differs from cocklebur in that leaves can only become induced by the environment and not by other leaves on the plant.

Obviously there is a great deal more to be done along these lines. Again it seems quite clear that the real problem is concerned with the biochemical nature of the flowering substance itself. We need to know how it is synthesized in the first place. What is its condition while it is being translocated? How is it maintained in the plant? Does it remain the same or is it changed in some way in the bud? How does it move into young cocklebur leaves, making them capable of inducing receptor plants? Why does this fail in Perilla? Why does it disappear in the cocklebur when there are no active buds?

Applied auxins or gibberellins, or the presence of very young leaves will retard or prevent this disappearance in the absence of active buds. In the last chapter we concluded that auxin in the *leaves* may destroy flowering hormone; now it appears that auxin in the *buds prevents* the destruction of flowering hormone or gives the bud the ability to produce it continually. Does this mean that auxin acts differently in the two tissues or that the hormone from the leaf is changed after it gets to the bud?

It should now be clear that flowering, in common with plant growth phenomena in general, is influenced to a very significant extent by growth regulators such as the auxins and gibberellins. One of the primary problems of future flowering research is to elucidate these influences. Some initial work has been concerned with simply trying to measure the growth regulator status of plants under different environmental conditions. The approach is to extract tissue with suitable solvents, and then to test the extracts in various bioassays. Growth promoter activity is often tested using oat coleoptile curvature or elongation tests, split pea stem tests, etc. Presence of gibberellins is indicated by effects upon dwarf mutants such as those of maize.

The approach has been followed by a number of workers, and the summary data shown in Table 10–2 were assembled by J. P. Nitsch at Gif-sur-Yvette (near Paris) based upon his experiments (29). It seems fairly clear from our previous discussions that the changes in auxin, gibberellins, etc., which accompany changes from short-day

TABLE 10–2. RELATIVE GROWTH REGULATOR LEVELS WITHIN PLANTS AS INFLUENCED BY ENVIRONMENTAL FACTORS

Long days	Short days
1. High auxins (Oat coleoptile promoters)	1. Low auxins
2. High gibberellins (dwarf maize bioassay), and Substance E.	2. Low gibberellins
3. High auxin synergists (primarily inhibitors of IAA-oxidase, such as alpha-tocopherol)	3. Low auxin synergists
4. High leucoanthocyanins (colorless)	4. High anthocyanins (colored)
5. Low inhibitors (coleoptile test)	5. High inhibitors

Vernalizing cold treatments

1. High auxins
2. High gibberellins
3. High sugars; in some plant families at least, fructose and fructosans.
4. High reducing substances such as glutathione
5. Low inhibitors (coleoptile tests)

Data from J. P. Nitsch (29).

to long-day conditions or vice versa are probably not in themselves responsible for the initiation of flowers. Yet these changes undoubtedly play important roles in development of the floral bud, elongation of flower stalks, and probably the status of the flowering hormone as discussed in this section.

THE CHANGE FROM MAKING LEAVES TO MAKING FLOWERS (8, 32, 37, 38)

We can now return to the exciting problem introduced in Chapter 1: How does the flowering hormone cause a change in direction of growth at the buds? The number of diverse and intricate forms found among living organisms is astronomical. It took biology a matter of two or three centuries just to describe these forms, and the process is still going on. What is responsible for the many differences in structure which are so evident to us? Virtually all living things begin their lives as a single cell. This cell divides, and the daughter cells produced at each division are usually almost identical. Such a process is marvellous in itself, but if this were the end of the story, only a mass of cells might be produced (which is what happens in the case of many tumors including cancer).

The other phase of the growth process is specialization, or differentiation, or morphogenesis. Cells specialize individually and in groups. Tissues become discernible, and soon we can see the organs which make up the organism itself. In the case of indeterminate plant growth this cell division and differentiation continues indefinitely. Cells at the tips of the stems remain embryonic, continuing to divide. Farther down the stem, cells are specializing into specific tissues such as pith, vascular elements, cortex, and epidermis. The epidermal and cortical cells further divide, producing small swellings which will eventually grow into leaves and axillary buds.

When the plant becomes reproductive, all of this changes. Leaves are no longer produced, but the reproductive organs begin to form. The flower parts appear to be modified leaves, but they are modified very extensively and their position on the stem is changed quite radically. The parts of the flower are all compressed together rather than being strung out at the nodes as are the leaves on caulescent plants. A new set of genes seems to be in control of things. All of this seems to occur in response to the arrival of the flowering hormone. How does it all work? How can a chemical entity (the flowering hormone) control the shape and structure of a three-dimensional organ such as a petal or stamen or pistil?

After years in which virtually no interest was shown in this fundamental phase of the flowering process, a few researchers are finally beginning to address themselves to the problem of what is happening in a morphological and in a biochemical way when the meristem is transformed from the vegetative to the reproductive condition.

1. *Work with Applied Antimetabolites*

As mentioned in the last chapter, the effects of certain anti-nucleic acid compounds seem to be in the bud rather than in the leaf. Zeevaart and James Bonner (38, 42) at Pasadena have shown that 5-FU is effective only when it is applied to the cocklebur bud during

P

the first 8 hr or so of the dark period. Thus it appears that 5-FU inhibits something which must be made in the bud even before flowering hormone is made in the leaf. They have further shown that RNA synthesis is inhibited. Thus the successful action of the flowering hormone when it arrives at the bud two or three days after induction may require conditions which are initiated by RNA synthesis at the very beginning of induction (see Chapter 9). Obviously, there is much to learn about all this.

Zeevaart (79) further found that the DNA inhibitor 5-fluorodeoxyuridine (5-FDU) would inhibit flowering of Japanese morning glory when it was applied to the bud, and that this effect could be fully reversed by application of DNA precursors such as thymidine. Thus DNA multiplication in the bud appears to be essential to flowering. The inhibitor is most effective only when applied near to the time of arrival of the hormone at the apex. If applied too early (40 hr) it is dissipated before arrival of the hormone and hence ineffective, and if it is applied after the bud has been transformed it is also ineffective. Microscopic examination has shown that cell division is inhibited by the 5-FDU. In this species 5-FU acts the same way as 5-FDU and not as it does in cocklebur.

Thus it appears that the flowering hormone must find dividing cells (multiplying DNA) if it is to turn on the genes for flowering. This agrees well with the observation in the last section that dormant buds do not respond to the stimulus. Does it also mean that multiplication of the stimulus (in cocklebur) requires multiplication of genetic material?

2. Vernalization, Juvenility, and the Induced State

The phenomenon of the induced state as described above seems to have a great deal in common with the condition induced by vernalization (Chapter 4) or with the transformation of a plant from the juvenile to the mature condition (Chapter 6). In all these cases two conditions clearly identifiable at their extremes, are involved, and the transformation is very stable and maintained in the plant through subsequent cell divisions of the growing tissue (with the exception of Perilla). Perhaps our best opportunity to study this phenomenon of a condition transferred through cell divisions is offered by the flowering process, in which the arrival of a hormone at the meristematic sites triggers the transformation to the induced state. Since the flowers will ultimately produce an embryo which is not in the induced state, a cytoplasmic change might seem more likely than a nuclear one. Yet the observation mentioned in the above section (that stability of the flowering hormone and hence the induced state probably depends upon the presence of actively dividing cells) might indicate the converse: the induced state is a condition of the nuclei of the meristems, caused by flowering hormone. Is this condition then reversed by formation of the gametes at miosis? At any rate it is easy to imagine that the change during vernalization is very similar. Is the change during maturation also similar? This could be a fruitful field for future investigation.

3. *The Morphology of Transformation*

It is somewhat surprising to learn that the morphological events taking place during transformation of the meristem have been observed only in a very limited number of investigations. The time of change from the vegetative to the reproductive form of growth can be pin-pointed almost within minutes and should provide an excellent subject for histological study of morphogenesis. There is a considerable background of information about the structure of plant meristems and the techniques required to study them. In recent years the field of histochemistry has developed rapidly so that it should be possible to study, in a preliminary way at least, the fundamental biochemistry of transforming cells at the meristem.

Figure 10–6 was drawn from photographs taken by W. F. Millington and E. L. Fisk at Wisconsin, and by Ralph Wetmore and his co-workers at Harvard University. Their study is the kind which should be done in considerable detail by many scientists around the world using many species until we at least have an accurate and complete descriptive idea of what goes on during floral differentiation. Wetmore summarized the sequence of events under the following five points: (1) The cells just below the "central zone" in the apical meristem were the first to become active in the differentiating bud. This activity consisted of cell divisions and some enlargement and was found in all species studied, as well as in the literature examined. This could be an important finding. Perhaps research efforts should be directed towards biochemical understanding of this phenomenon. (2) Activity then spread to the other cells, including those in the central zone. Cells in this zone became smaller by dividing, and

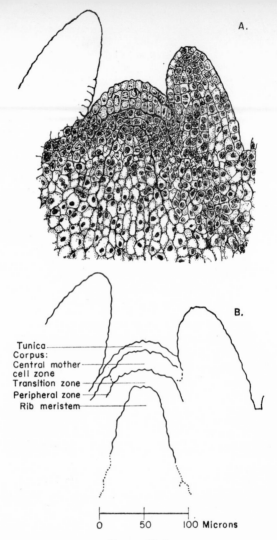

A.

B.

Tunica
Corpus:
Central mother
cell zone
Transition zone
Peripheral zone
Rib meristem

0   50   100 Microns

FIGURE 10-6

Histological changes during transition from the vegetative to the
reproductive condition. A. Median longitudinal section through the
shoot tip of a vegetative cocklebur plant. Drawn from a photograph in
W. F. Millington and E. L. Fisk, 1956, *Amer. J. Botany* 43, 655–665.
B. Tracing of the section in A, showing the zonation proposed by
Millington and Fisk. Note in A and B the large, lightly staining nuclei
in the tunica; the 2 rows of cells almost continuous with the tunica, also
having large, lightly staining nuclei (the central mother cell zone); the
2 to 3 rows of cells immediately below, having much denser nuclei (the
transition zone); and finally the rib meristem, having cells at the top
similar to the transition zone cells, but cells with larger vacuoles and
smaller nuclei below. C. Beginning the transition to the floral bud (an

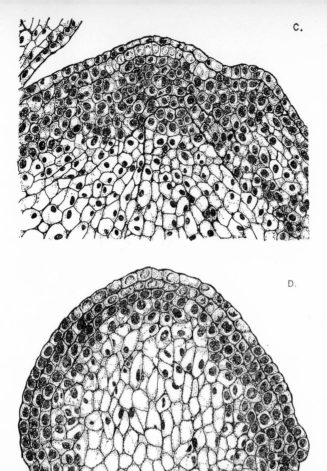

C.

D.

early Floral Stage 1 — see Fig. 5–8). Only one row of mother zone cells is obvious, but these and the tunica still have large, relatively clear nuclei. There are now 3 to 6 rows of cells in the transition zone, the primary site of activity at this early stage. D. Floral Stage 3, clearly a floral bud (inflorescence primordium). The most striking event has been the enlargement of the rib meristem cells along with some of the transition cells, beginning to produce the characteristic shape of the cocklebur staminate floral head. The cells with light nuclei in the tunica and mother cell zone are still evident, but they will gradually disappear as the bud develops. C and D were drawn to the scale of A and B from photographs in R. H. Wetmore, E. M. Gifford, Jr., and M. C. Green, 1959, in R. B. Withrow (editor) *Photoperiodism and Related Phenomena in Plants and Animals*, 255–273, American Association for the Advancement of Science, Washington, D.C.

eventually there was a layer of meristematic cells 2 or 3 cells deep covering the entire bud. The bud at this stage is an easily recognizable floral primordium and has passed far beyond the initial changeover from the vegetative to the reproductive state. (3) Enlargement of the pith rib meristem cells caused swelling of the bud. (4) The layer of meristematic cells covering the surface formed the bracts, flower parts, etc. (5) Extension of the axis stopped, and apical dominance is lost.

An exciting recent report concerns the preliminary findings of E. M. Gifford and H. B. Tepper at Davis, California (see 37). They have initiated histochemical studies on the developing meristem of pigweed (*Chenopodium album*) and other species. So far they have observed sharp increases in RNA and in protein in the cytoplasm of meristematic cells at the time of transformation to the reproductive condition. It would appear that the genes are busily synthesizing RNA at this time, and that the RNA is in turn causing an increase in enzymes (protein), according to our current concepts of protein synthesis as controlled by messenger RNA from the nucleus. This is the sort of thing one might expect. Transformation should consist of the formation of many new enzymes which are required for floral development.

Most interesting of all, Gifford and Tepper observed in the nuclei of transforming cells a sharp decrease of the basic protein histone. R. C. Huang and Bonner (52) at Pasadena have recently shown with an *in vitro* system that DNA fully complexed with histone is completely inactive in synthesizing RNA. Thus it appears that turning on a gene requires removal of histone, which in turn results in production of RNA and then the enzyme controlled by the gene. The observations of Gifford and Tepper seem to indicate that floral transformation consists of turning on nearly *all* the genes by removing nearly *all* the histone, which leads to the synthesis of much RNA and many enzymes. Later the histone begins to reappear, indicating perhaps that a more select set of genes is now in control of the situation. Is our flowering hormone simply a remover of meristematic histone?

It would seem that we are on the threshold of some exciting discoveries. Such important steps forward could make parts of this discussion obsolete before it can appear in print. We can await rapid advances.

The Present and The Future

In the field of classification of the many response types, there is still a long way to go. Almost any species can be studied in just a little more detail to yield considerable information about responses to changing temperatures, special light qualities, etc. It may be many years before we can properly evaluate the relationship between biological diversity and uniformity. There is also much to learn about the ecological aspects of the flowering process. Based upon work on the many response types, we should be able to gain more and more understanding concerning the specific adaptations of the members of a plant community.

In physiological fields, we are advancing rapidly. We seem to be approaching some sort of understanding of the response to cold. Work with the gibberellins and other extracts should lead in the near future to improved concepts of how plants respond to cold by being converted to the induced state. Advances in biochemistry may take us ever closer to an understanding of the preparation for response to photoperiod and of the metabolic pathways involved in synthesis of the flowering hormone. Extraction and characterization of phytochrome is surely one of the most significant steps in recent years. If present preliminary reports on the extraction of the flowering hormone bear fruit, we might expect rapid advances in our overall understanding of the biochemistry of the flowering process, although the nature of the induced state may prove to be somewhat refractory. The work summarized above might even indicate that the fundamental problem of transformation of the meristems could yield to solution in the forseeable future. We at least have some reasonable biochemical ideas about these fields.

Perhaps the most perplexing frontier at present concerns time measurement. Discovery of phytochrome and work on other possible pigment systems form necessary background, but at this moment the real nature of timing seems to have completely eluded us. New concepts and approaches are needed in this field. Perhaps the proven methods of environmental manipulations with whole plants and finally extractions for biochemical study are unequal to the task of understanding timing. New concepts of both methodology and life function may have to be developed.

We live in exciting times. There is much left to learn about biology and the flowering process, and thus there is ample challenge for the inquiring mind. At the same time, a survey of our accomplishments also yields a fairly satisfying picture. We can at least formulate a coherent and reasonable outline of the flowering process. It is easy to be impressed by the many ways that plants respond to their environment, the low temperature, short- and long-day responses, etc. There is a great diversity but some underlying uniformities are becoming apparent. Control of plant processes by phytochrome offers our first real insight into the photo- and biochemical interactions of a plant with the low intensity light environment. The discovery of a time measurement is itself most impressive, even if we still fail to understand the mechanisms involved. There is much to learn, but at least we have developed a concept of biochemical synthesis of a flowering stimulus which will control the transformation of the meristem to the reproductive condition. We are even beginning to gain some insight into how this transformation might take place.

APPENDIX

Continuing work on the following list has impressed me deeply with its highly tentative nature. Complete reliance cannot be placed on the majority of species as listed. Thus the list has only one real value: to demonstrate to the reader who is first encountering the physiology of flowering, that there is a fantastic diversity of response among the many species which have so far been studied. This is an important conclusion. It indicates the complexity of the flowering process, the number of genetically controlled steps which must be directly or indirectly involved, and it implies an intricate and complex ecology of flowering.

There are at least three reasons why all but the most thoroughly studied, completely standardized genetic types in the following list must be regarded with some suspicion.

1. Studies are incomplete. Obviously, there are so many possible simple combinations of day-length and temperature, not to mention changes in day-length and temperature with time, changes in light intensity, light quality, humidity (flowering of one plant is now thought to be controlled by humidity! — P. Chouard, personal communication), etc., that few if any plants have ever been subjected to all of them. Thus future work will probably move many plants to more complex categories.

2. The genetic variability within species and varieties cannot be overemphasized. In a few cases (e.g. *Chrysanthemum*) the various responses found within a single species are indicated in the following list.

3. In many cases where work is adequate and varieties standardized, problems of classification still arise because a response may be somewhat border-line in nature. This is illustrated in Fig. A–1 for the long-day response. If the experimenter feels that he has waited an "infinite time" (!), then he can feel secure about assignment to an absolute-long-day category. Obviously there are many possible degrees of a quantitative response, and it may be especially difficult

Q

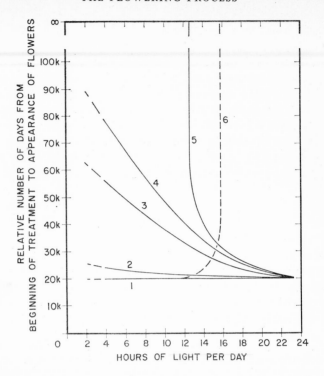

FIGURE A-1

Flowering response to different day-lengths, plotted as time required for appearance of flowers. Curves are not based on actual data (except for number 6) but are illustrative of the kinds of responses which might be expected. K following the numbers on the ordinate indicates that these numbers might be multiplied by a constant, and are thus arbitrary. The 20K time for the earliest appearance of flowers is also arbitrary. Curve 1 is a true day-neutral plant; curve 2 would probably also be classed as a day-neutral plant, although it is slightly promoted by long days. Curves 3 and 4 represent quantitative long-day plants, but 4 is more sensitive than 3. Curve 5 is a true absolute long-day plant. Curve 6 is a true absolute short-day plant (the cocklebur).

to firmly define the day-neutral response (curve 1 or curve 2?). To complicate matters even more, a plant may be completely day-neutral when days-to-flowering are considered as in Fig. A–1, but if the number of flowers per plant is measured instead, then there may be a strong promotion by long days (e.g. *Leucanthemum cebennense* and *Saxifraga rotundifolia*).

Thus the list may be considered in the following light: the plants shown have at one time or another displayed the responses indicated by the category in which they are now placed. An investigator might . expect them to respond again in a like manner, but he should not be greatly surprised if they don't. His conditions may be different, he may have plants with a different genetic constitution, and his ideas of classification may differ from those of the original investigator.

Except for a few plants for which references are directly cited, most plants were taken from two lists. Those followed by H and a number are taken from: William S. Spector (editor, 1956, *Handbook of Biological Data*, W. B. Saunders Co., Philadelphia and London, Table 391, page 460). The number refers to the number of the plant in the table. Plants marked with C are taken from P. Chouard, 1960, Vernalization and its relations to dormancy, *Annual Review of Plant Physiology*, 11, 191–238. I am grateful to Prof. Chouard for examining the list (November, 1962) and offering suggestions (he strongly emphasized the genetic variability which might be expected). I am also indebted to Mrs. Catherine Kline who assembled the data on punched cards and helped with compilation of the lists, especially the common names.

−• 1. **Day-neutral plants, no causative temperature effect.** These are the plants with the least response to their environment as far as their flowering goes. They flower at about the same time under virtually all conditions. Probably most of the plants in this category are valid, but some may be moved after future research to categories such as: promotion of flowering by alternation of temperature. The tomato plant is a good example of a plant that has already been changed in this way.

Cucumis sativus, H. 9	Cucumber
Euphorbia peplus, C. 218	
Fagopyrum tataricum, Skok and Scully, 1955. *Botan. Gaz.* 117: 134–141	Buckwheat
Fragaria chiloensis, H. 19	Strawberry, everbearing
Gardenia jasminoides fort., H. 84	Cape jasmine
Gomphrina globosa (personal observation)	Globe-amaranth
Gossypium hirsutum, H. 59	Cotton, one variety
Helianthus tuberosus, H. 1	An artichoke variety
Ilex aquifolium, H. 86	English holly
Impatiens balsamina, H. 71	Balsam
Lunaria annua, C. 218	Honesty
Nicotiana Tabacum, H. 64	A tobacco variety
Nicotiana silvestris, C. 198	
Phaseolus lunatus, H. 2	Lima bean, a variety
Phaseolus vulgaris, H. 3	A variety of String bean
Poa annua, H. 27	Bluegrass, annual

Rhododendron sp., H. 70	Azalea, coral bell
Scrofularia peregrina, C. 218	
Senecio vulgaris, C. 218	
Solanum tuberosum, H. 15	A variety of potato
Viburnum spp. Downs, R. J., A. A.	
Piringer, 1958, *Proc. Amer. Soc. Hort.*	
Sci. 72: 511–513	
Zea mays, H. 33	Maize or corn

! ●   2. **Day-neutral, quantitative promotion by low temperature.** These plants will flower under virtually any conditions, but they flower sooner if they receive a low temperature treatment.

Allium cepa, H. 12 (non-inductive response)	A variety of onion
Lathyrus odoratus ? C. 207	Sweet pea
Lens culinaris ?, C. 207	Lentil
Pisum sativum, H. 13, C. 207, 218	Garden pea variety
Pelargonium hortorum, H. 85	Fish geranium
Vicia Faba ? C. 207	Broad bean
Vicia sativa ? C. 207	Vetch
Viola tricolor ? H. 93	Pansy

↑ ●   3. **Day-neutral, promotion by high temperature.**

Fuchsia hybrida, H. 83	Fuchsia
Oryza sativa, H. 38	Summer rice

↑ ●   4. **Day-neutral, promotion by temperature alternation (Thermoperiodism).**

Capsicum frutescens, H. 14	Pepper
Lycopersicon esculentum, H. 21, C. 206;	Tomato
Fig. 1–A	

↓ ●   5. **Day-neutral, low temperature required.** (Vernalization). Members of this category which can be placed here without any question refute a long-held idea that all cold-requiring plants also required long-days. It may still be true that all cold-requiring plants which respond as seeds or seedlings require long days. The plants listed here must all reach some stage of vegetative growth before they will respond to cold. A few species in which the day-length requirement is in doubt are marked by a question mark.

Agrimonia eupatoria, C. 212	Agrimony
Apium graveolens, II. 8, C. 203	Celery
Arabidopsis thaliana ? C. 200	Variety "Stockholm"
Cardamine amara, C. 212	Bittercress
Centaurium minus ? C. 222	Centaury
Draba aizoides, C. 211, 218	Whitlow-grass
Draba hispanica, C. 211	
Eryngium variifolium C. 212, 222	Eryngn
Erysimum spp., C. 212, 218	Wallflower
Euphorbia lathyris, C. 205, 222	Caper spurge
Geum urbanum, C. 208, 210, 218	
,, *bulgaricum*	
,, *intermedium*	
,, *canadense*	
,, *album*	
,, *macrophyllum*	

Hydrangea macrophylla ? H. 87   Hydrangea
Lunaria biennis, C. 218, 236, 206
Lychnis coronaria ? C. 212   Dusty miller, variety
    „   *viscaria* ?
    „   *flos-cuculi* ?   Ragged robin
Pyrethrum cinerariaefolium, C. 213   Dalmation pyrethrum
Saxifraga rotundifolia, C. 211, 218, 222
Scrofularia alata, C. 212, 222   Figwort
    „   *vernalis*, C. 204, 222
Senecio Jacobea, C. 205   Groundsel

-◉ 6. **Quantitative short-day plants; no causative temperature effect.**
Andropogon virginicus, H. 30   Broomsedge
Cannabis sativa, Borthwick and Scully, Hemp, variety Kentucky
    1954, *Botan. Gaz.* 116: 14–29
Chrysanthemum hortorum or morifolium
    (hybrids), C. 214, 218
Cosmos bipinnatus, H. 279   Cosmos
Cucurbita sp., Fig. 1–H.   Squash
Datura stramonium, H. 111   Jimson weed
Gossypium hirsutum, H. 59   Cotton, a variety
Helianthus tuberosus, H. 1   Artichoke, a variety
Saccharum officinarum, H. 47   Sugar cane
Senecio cruentus, H. 77   Cineraria
Solanum tuberosum, H. 15   Potato, 2nd variety
Zinnia sp., Fig. 1–G

-•◉ 7. **Quantitative short-day plant at high temperature; day-neutral at low temperature. No causative effect of temperature.**
Holcus sudanensis, H. 46   Sudan-grass
Malva verticillata, H. 113   Mallow
Salvia splendens, H. 97   Scarlet sage
Zygocactus truncatus, H. 74   Crab cactus

↕◉ 8. **Quantitative short-day plants promoted by low temperature.**
Allium cepa, H. 12   Onion, 2nd variety (may
                                        be another example of a
                                        non-inductive effect in a
                                        cold promoted species).

↑◉ 9. **Quantitative short day promoted by high temperature.**
Amaranthus graecizans, H. 118   Tumbleweed
Chrysanthemum hortorum C. p. 214, 218 Variety 2

↑◉ 10. **Quantitative short-day plant promoted by high temperature; critical dark period inversely proportional to high temperature.**
Glycine soja, H. 62   Soybean, Mandell

↕◉ 11. **Quantitative short-day plant promoted by temperature alternation.**
Capsicum frutescens, H. 14   Pepper, second variety
Chrysanthemum spp, Schwabe, W. W.,   Variety 8
    1957. *J. Exptl. Botany* 8: 220–234
Lycopersicon esculentum, H. 21   Tomato, 3rd variety

↓◉ 12. **Quantitative short-day plants; low temperature required.**
Chrysanthemum hortorum, C. 214, 218   Variety 3

-◯   13. **Quantitative long-day plants; no direct temperature effect.**
Brassica rapa, H. 22   Turnip
Dianthus prolifer, C. 218
Fragaria chiloensis, H. 19   A variety of everbearing
                                            strawberry
Hordeum vulgare, H. 23, C. 197, Fig. 1-J   Spring barley
Nicotiana Tabacum, H. 65   Havana tobacco
Nigella arvensis, C. 218   Fennel-flower
Nigella damascena, C. 218   Love-in-a-mist
Oenothera rosea, C. 218   Evening primrose
Pisum sativum, H. 13   Garden pea, 2nd variety
Scrofularia arguta, C. 218
Secale cereale, H. 40, C. 193   Spring rye
Solanum tuberosum, H. 15   A variety of potato
Sonchus oleraceus, H. 117   Sowthistle
Sorghum vulgare, H. 45   *Sorghum*
Triticum aestivum, H. 51, C. 197   Spring wheat

-•◯   14. **Quantitative long-day plant at high temperature; day-neutral at low temperature. No causative effect of temperature.**
Anthemus cotula, H. 107   Dog fennel
Antirrhinum majus, H. 99   Snapdragon
Begonia semperflorens, H. 72   Begonia
Centaurea cyanus, H. 78   Cornflower
Matthiola incana, H. 100, C. 206   German stock
Medicago sativa, H. 54   Alfalfa variety
Petunia hybrida, H. 94, C. 198   Petunia
Poa pratensis, H. 28   Kentucky bluegrass
Solanum nigrum, H. 114   Nightshade
Vicia sativa, H. 67   Spring vetch

↕◯   15. **Quantitative long-day plants promoted by low temperature.**
Agrostemma githago, C. 218   Corncockle
Allium cepa ? H. 12   Onion variety
Cichorium endivia, C. 201   Endive
Lactuca sativa, H. 11, C. 200   Lettuce
Oenothera strigosa, H. 115   Evening primrose
Sinapis alba (*Brassica hirta*), C. 218, 206   White mustard
Trifolium spp., H. 57   Clover

↑◯   16. **Quantitative long-day plants promoted by high temperature.**
Callistephus chinensis, H. 69   China aster

↕◯   17. **Quantitative long-day plants promoted by temperature alternation.**
Lycopersicon esculentum, H. 21   Tomato, 2nd variety

↓◯   18. **Quantitative long-day plants: low temperature required.**
Campanula persicaefolia, C. 204
      „   *alliariaefolia*
      „   *primulaefolia*
Cheiranthus cheiri, C. 213, 218   Wallflower
Cynosurus cristatus, C. 211   Dogtail
Daucus carota, H. 7, C. 203   Carrot
Dianthus barbatus   Sweet William

Digitalis purpurea, H. 82, C. 204, 218   Foxglove
Iberis intermedia, C. 205, 218   Candytuft, variety
   Durandii
Leucanthemum cobennense, C. 213, 218   Daisy
Lychnis coronaria, C. 212   Dusty miller, variety
Scabiosa sanescens, C. 212, 218
   „ *succisa*, C. 210   Devil's-bit
Teucrium scorodnia, C. 213, 218, 222   Germander

19. **Quantitative long-day plants: high temperature required.**
Camellia japonica, Bonner, J. 1947,   Camellia
   Amer. Soc. Hort. Sci. 50: 401–408

20. **Short-day plants (qualitative or absolute); no direct temperature effect.**
Ambrosia elatior, H. 116   Ragweed
Andropogon gerardi, H. 25   Beardgrass
Bryophyllum pinnatum, H. 73   Bryophyllum
Cattleya trianae, H. 92   Orchid
Chenopodium album, H. 112, Fig. 1–E,   Pigweed varieties
   3–8
Chenopodium rubrum, Cumming, B. G.,   Pigweed
   1959, *Nature*, 184: 1044–1045
Chrysanthemum morifolium, C. 214, 218   Variety 4
Coffea arabica, Piringer, A. A. and H. A.   Coffee
   Borthwick, 1955, *Turrialba* 5: 72–77
Ipomoea batatas, H. 20   Sweet potato
Ipomoea hederacea, H. 90   Morning glory
Kalanchoë blossfeldiana, H. 88   Kalanchoë
Lemna perpusilla, 6746, Hillman, W.   Duckweed
   Amer. J. Bot. 46: 466–473, 1959
Lespedeza stipulacea, H. 60   Bush clover
Perilla ocymoides, Fig. 1–F
Phaseolus lunatus, H. 2   Lima bean, 2nd variety
Phaseolus vulgaris, H. 3   String bean
Solidago spp., H. 108   Golden rod
Zea mays, H. 33   Maize or corn, 2nd
   variety

21. **Short-day plants; no causative temperature effect; critical dark period inversely proportional to temperature.** Demonstration of the inverse relationship between critical dark period and temperature requires a fairly elaborate physical facility for experimentation. Thus only three plants can be listed in this category at present, although it seems quite likely that further experimentation would move a number of plants from the above category into this one. The cocklebur is included in this category.
Chrysanthemum indicum, H. 76   Chrysanthemum
Fragaria chiloensis, H. 18   Strawberry, 1st variety
Xanthium pennsylvanicum, H. 103,   Cocklebur
   Fig. 1–B

22. **Short-day plants at low temperature; day-neutral at high; no causative effect of temperature on flowering.**
Cosmos sulphureus, H. 81   Cosmos, orange flare

23. **Short-day plants at high temperature; day-neutral at low: no direct effect.** It is interesting that the last two examples should fall in the same category. Maryland Mammoth tobacco was the first plant shown to have a short-day requirement, and a study of the flowering of this plant led to the discovery of photoperiodism. The Japanese morning glory is one of the newest plants studied by plant physiologists, and yet it has been studied almost more extensively than the tobacco, or any other plant for that matter.

Chenopodium album, Fig. 3–8	Pigweed variety
Nicotiana Tabacum, H. 66, C. 198	Maryland Mammoth tobacco
Pharbitis Nil, Ogawa, Y., 1960, *Bot. Mag.* (*Tokyo*), 73: 334–335; Takimoto, A., Tashima, Y., and Imamura, S., 1960, *Bot. Mag.* (*Tokyo*), 73: 377, Fig. 1–C	Japanese morning glory, var. violet

24. **Short-day plants at high temperature; long-day plants at low; no causative effect.** Such a complete change in response type provides a very striking category.

Euphorbia pulcherrima, H. 96	Poinsetta
Ipomoea purpurea, H. 91	Morning glory

25. **Short-day plants promoted by high temperature.**

Cosmos sulphureus, H. 80	Klondike cosmos
Chrysanthemum morifolium, C. 214, 218	Variety 5
Oryza sativa, H. 39	Winter rice

26. **Short-day plants promoted by high temperature; critical dark period inversely proportional to temperature.**

Glycine soja, H. 61, Fig. 1–D	Soybean, Biloxi
Viola papilionacea ? H. 102	Violet

27. **Short-day plants; require low temperature.** This is also a significant category, since many workers have felt that only long-day plants might have a low temperature requirement.

Chrysanthemum morifolium, C. 214, 218	Variety 6

28. **Long-day plants; no direct effect of temperature.** It is quite likely that a number of these might display a temperature interaction if investigated in the proper way.

Agropyron smithii, H. 53	Wheatgrass
Agrostis nebulosa, H. 32	Cloudgrass
Agrostis palustris, H. 26	Bentgrass
Alopecurus pratensis, H. 35	Foxtail
Anagallis arvensis, C. 218	Pimpernel
Anethum graveolens, H. 10, C. 203?	Dill
Avena sativa, H. 36 C. 197	Oat
Chrysanthemum frutescens, H. 75	Paris daisy
Chrysanthemum Leuchanthemum, H. 106	Ox-eye daisy
Dianthus superbus, C. 212, 218	Carnation
Festuca elatior, H. 34	Fescue
Hibiscus syriacus, H. 68	Althea
Lolium temulentum, Evans, L. T., 1958, *Nature*, 182: 197–198	Ryegrass (induced by a single cycle)

Melilotus alba, H. 63	Sweetclover
Mentha piperita, var. *vulgaris*, Langston and Leopold. *Proc. Amer. Soc. for Hort. Sci.* 63: 347–352, 1954	Peppermint
Oenothera acaulis, C. 218	
Phleum nodosum, H. 50	Pasture timothy
Phleum pratensis, H. 49	Hay timothy
Phalaris arundinacea, H. 31	Canary-grass
Raphanus sativus, H. 16, Fig. 1–L	Radish
Ricinus, spp., Scully and Domingo, 1947, *Botan. Gaz.* 108: 556–570	Castor-bean, variety Kentucky 38
Rudbeckia hirta, H. 105	Coneflower
Scabiosa ukranica, C. 218, 222	
Sedum spectabile, H. 98	Sedum
Spinacia oleracea, C. 201, 218, Fig. 1–K	Spinach, variety 4
Trifolium spp., H. 56	Clover species
Trifolium pratense, H. 58	Red clover, 2nd variety

-◯⸱   29. **Long-day plants; no causative effect of temperature; critical dark period inversely proportional to temperature.** The one example is a classical object for photoperiodism research.

> *Hyoscyamus niger*, H. 109, C. 192, 197,   Henbane, annual strain
> Fig. 1–I

-◯◖◗   30. **Long-day plants at low temperature; quantitative long-day plants at high temperature; no causative effect of temperature.**

> *Beta vulgaris*, H.4   Garden beet
> *Brassica pekinensis*, H.5   Chinese cabbage

-◯•   31. **Long-day plants at low temperature; day-neutral at high; no causative effect of temperature.**

> *Delphinium cultorum*, H. 89   Larkspur

-•◯   32. **Long-day plants at high temperature; day-neutral plants at low temperature; no causative effect of temperature.**

> *Cichorium intybus*, H. 6   Chicory

-⬦   33. **Long-day plants; no causative temperature effect; low temperature will replace the long-day requirement.** This differs from No. 32, in that the low temperature treatment is too low for active growth, and hence the effect is inductive. This is a very interesting category.

> *Trifolium subterraneum*, Morley, F. H.   Subterraneum clover
> and L. T. Evans, 1959. *Australian J.*
> *Agric. Research* 10: 17–26

-⬦   34. **Long-day plants; no causative effect of temperature; high temperature will replace a long-day requirement.** This is equally interesting, but it might be interpreted as an example of No. 31.

> *Rudbeckia bicolor*, H. 104   Coneflower

-◯⼁•   35. **Long-day plants; no direct temperature effect; low temperature induces the day-neutral response.** This differs from No. 33 because the low temperature treatment is applied to moist seeds, rather than seedlings or young plants.

> *Spinacia oleracea*, C. 201   Spinach, variety "Nobel"

↓○ 36. **Long-day plants promoted by low temperature.**

Avena sativa, H. 36, C. 197	Oat
Bromus inermis, H. 29	Bromegrass
Dianthus arenarius, C. 213, 218	Carnation
Dianthus gallicus, C. 212, 218	Carnation
Hordeum vulgare, H. 24, C. 197	Winter barley
Lolium italicum, H. 42	Italian ryegrass
Oenothera suaveolens, C. 203, 218	Evening primrose
„ *longiflora*	
„ *stricta*	
Spinacia oleracea, C. 201	Spinach, variety 3
Triticum aestivum, H. 52, C. 197	Winter wheat, most varieties

↑○ 37. **Long-day plants promoted by high temperature.**

Phlox paniculata, H. 95	Phlox

↓○ 38. **Long-day plants with a low temperature requirement.**

Anagallis tenella, C. 214, 218	Pimpernel
Beta vulgaris, H. 55, C. 199	Sugar beet
Cichorium intybus ? C. 204	Chicory, variety
Crepis biennis ? C. 205	Hawksbeard
Dianthus coesius, C. 213, 218	Carnation
Dianthus graniticus, C. 213, 218	Carnation
Lolium perenne, H. 43, C. 211	Early perennial ryegrass
Lolium perenne, H. 44, C. 211	Late perennial ryegrass
Lysimachia nemorum, C. 214	Loosestrife
Oenethera lamarckiana, C. 202, 218	Evening primrose from forests of Fontainebleau
Oenothera parviflora biennis, C. 222, 218, 202	Evening primrose
Oenothera strigosa biennis, C. 202, 236, 218	Evening primrose
Saxifraga hynoides, C. 214	
Spinacia oleracea, H. 17	Spinach, variety 1

↓○⁻ 39. **Long-day plants; low temperature required; critical dark period is inversely proportional.** This variety of Hyoscyamus is also a classic in research on photoperiodism and vernalization.

Hyoscyamus niger, H. 110, C. 192, 197 218	Henbane, biennial strain

-◑ 40. **Quantitative short-long-day plant; short-day effect replaced by low temperature; no direct temperature effect.** This includes the classic winter rye, studied so extensively over a period of many years. According to early studies, this plant would have been classified only as a quantitative long-day plant, promoted by low-temperature (category No. 15). Detailed studies, however, warrant placing of the plant in this complex category, and the possibility is immediately raised that many other plants might find themselves in such complex categories if study of them were carried out to a sufficient degree of detail.

Secale cereale, H. 41, C. 194	Winter rye
Iberis durandii, C. 223	Candytuft

↓◖ 41. **Absolute short quantitative long-day plants which require low temperature**

    Poa pratensis, Peterson, Maurice L. and   Kentucky bluegrass,
    Loomis, W. E., 1948. *Plant Physiol.*   variety 2
    24: 31–43

-◖ 42. **Quantitative long-short-day plants; no direct temperature effect.**

    Chrysanthemum ssp., C. 214   Variety 7

-◖ 43. **Quantitative long-short-day plants; no direct temperature effect; long-day effect quantitatively replaced by low temperature.**

    Chrysanthemum spp., C. 214   Variety 9

-● 44. **Intermediate-day plants; no direct temperature effect.** Sugar cane is now thought by many researchers to be a short-day plant instead of an intermediate-day plant.

    Chenopodium album, Fig. 3–8   Pigweed varieties
    Tephrosia candida, H. 101   Hoary pea
    Saccharum officinarum, H. 48   Sugar cane,
                var. 28NG 292

-◕ 45. **Plants quantitatively inhibited by intermediate day lengths, no direct effect of temperature.** The discoverers of this interesting response call it ambiphotoperiodism.

    Madia elegans, C. Ch. Mathon et M.
    Stroun. *Con.-Troisieme Congress*
    International de Photobiologie, 1960,
    Copenhagen.

◐ 46. **Short-long-day plants; no direct temperature effect; short-day replaced by low temperature.** Chouard calls the replacement of short-day requirement by low temperatures "Wellensiek's Phenomenon", after its discoverer. Chouard (personal communication) says a dozen or more other species are now known for this category. He also mentioned the interesting discoveries that one variety of *Scabiosa pratensis* requires *either* low temperature *or* short days, after which it is completely day-neutral, and that several strains of *Lolium perenne* require short days, chilling, and long days in that order, with no apparent replacement or interactions.

    Campanula medium, C. 201, 218   Canterbury bells

↓◐ 47. **Short-long-day plants; low temperature required.**

    Dactylis glomerata, H. 37, C. 211.   Orchard grass
    Gardner, F. P. and Loomis, W. E.,
    1953, *Plant Physiol.* 28: 201–217

-◑ 48. **Long-short-day plants; no direct temperature effect.**

    Bryophyllum daigremontianum, Resende,
    F., 1952. *Portugaliae Acat. Biologica.*
    Series A–III: 318–322.
    Cestrum nocturnum, Sachs, R. M., 1956.   Night-blooming jasmine.
    Plant Physiol. 31: 430–433.

NOTE ADDED IN PROOF

As a by-product of visits to controlled environment facilities, I have had the opportunity during the spring of 1963 to visit a number of scientists working on the physiology of flowering in Europe, Great Britain, and Russia. Although their results might be considered mostly preliminary, they provide such a good insight into what might be expected in the near future in this field, that I am unable to refrain from mentioning a few examples.

A number of workers, especially W. W. Schwabe at Wye College in England, are experimenting with new plants which might provide novel approaches to the flowering process. S. J. Wellensiek and J. Doorenbos at Wageningen in Holland have an active research group. One of their students has evidence that the long-day inhibitor (see pages 156-8) is a translocatable substance. In addition, they are quite convinced that response to cold in vernalization requires the presence of dividing cells (pp. 52, 194, and 198). Workers at Imperial College in London also hope to look into this question. They are further interested in nucleic acid metabolism as it relates to induction. This problem, as well as the role of nucleic acids in transformation of the bud, is being actively investigated at a number of laboratories. Cellular nucleic acid metabolism is being studied at Liège in Belgium (where I was also very impressed with photomicrographs of transforming meristems), and J. Heslop-Harrison at Birmingham in England has interesting microanalytical results with transforming buds. R. G. Butenko, working in Chailakhyan's laboratory in Moscow, has been able to induce flowering in isolated stem tips by application of kinetin or a mixture of RNA nucleotides.

A number of laboratories have promising results with extractions. We can hope for breakthroughs in this area. Extractions made in Chailakhyan's laboratory have provided further evidence that a product of long-day treatment is gibberellin (pp. 60-1). Chailakhyan's two-hormone theory seems to be on a better footing than is implied by my mention of it on p. 61. He points out that only the evidence for the short-day component still lacks conformation by extraction.

Discussion with G. Meijer at Eindhoven in Holland convinced me that I have previously interpreted his results incorrectly (see p. 123). Rather than contradicting the results of W. Könitz, his experiments collectively seem to support the idea that far-red light will act in an inhibitory way when it is given during the main light period (see also below).

Many of the experiments listed as being unpublished will appear in three technical papers in the journal *Planta* almost concurrently with the publication of this book. These include the experiment of Figs. 3–7, 7–8, 8–5, 8–6, 9–2, 9–5, 9–6, 9–7, 9–8, 9–13, and 9–15, Table 9–1, and the cobaltous ion results mentioned on pp. 142–3.

Some of the experiments which we have been performing in recent months on Timing and the High Intensity Light Process provide answers to certain of the questions raised in the text. To begin with, the inhibitory effect of a red light interruption given 8 hr after the beginning of the dark period cannot be reversed by application of sucrose, even though this interruption is followed by 12 hr or more of light (see pp. 117, 135, and 142). Thus, although ample sugar is present, the clock which measures the dark period will not restart after such an interruption. Why, then, did the experiment of Liverman and Bonner (p. 98) succeed? Probably because they gave 24 hr of flashing light, which might allow the clock to make a complete cycle and be back at its starting position, after which the sucrose depleted during the 24 hr of flashing light might indeed be limiting. We have tested this explanation with one preliminary experiment. Applied sucrose essentially failed to make a long dark period effective after only 8 hr of flashing light but partially succeeded after 16 hr of flashing light and succeeded quite well after 24 hr of flashing light! This is strong evidence in favor of an oscillating timer which times the dark period in the flowering process of cocklebur.

We have wondered how much light might be required to start the clock following a light interruption 8 hr after beginning of the dark period. To find out, we exposed cocklebur plants to 8 hr dark (or Japanese morning glory to 10 hr dark — the critical night length), then to different durations of light in the growth chamber, and then to a long inductive dark period (12 hr with cocklebur, 18 hr with Japanese morning glory). For the first hour or two, flowering is inhibited, but after 8 hr of light (both plants) the level of flowering reaches that of uninterrupted controls. Since sucrose fails to replace this requirement for 8 hr or more of light, something in addition to photosynthesis must be involved. Furthermore, although the effect is complicated, the 8 hr requirement seems to be independent of intensity. (Subsequent level of flowering is intensity dependent, indicating a photosynthesis *component*.) One experiment has been performed to determine the action spectrum for clock restarting. Using Japanese morning glory, plants were given 10 hr darkness, 1 hr white light (to inhibit flowering), 10 hr under various light qualities at equal intensities using the monochrometer installation at Tübingen), and then 18 hr darkness. The results of this experiment provide a typical phytochrome action spectrum for restarting the clock! Red light was most effective, orange nearly as effective, blue much less, and green and far-red ineffective.

As in Chapter 6, we may conclude that photosynthesis is required to provide substrates for the dark period reactions, but then photosynthesis is ultimately necessary for growth of the plant! In addition an essential part of the High Intensity Light Process must be a "winding" of the clock which will measure the length of the dark period. This process requires presence of F-phytochrome for a time interval of about 8 hr. This also agrees with the observations of Könitz (p. 123), and it may be interpreted as being in agreement with the theory of Bünning (pp. 133–6), which says that light (F-phytochrome) promotes during one part of the 24 hr cycle and inhibits during the other. It may even provide some understanding of this theory.

Tübingen
May 25, 1963

SELECTED BIBLIOGRAPHY AND LITERATURE CITED BOOKS

1. *Biological Clocks*, 1960. Cold Spring Harbor Symposia on Quantitative Biology, XXV. Cold Spring Harbor, L.I., New York: Long Island Biological Association, 524 pp.
2. BÜNNING, E. 1963. *Die Physiologische Uhr*. Springer-Verlag, Berlin-Göttingen-Heidelberg. 153 pp. (English edition in press.).
3. HILLMAN, WM. S. 1962. *The Physiology of Flowering*. Holt, Rinehart, & Winston, New York, 164 pp.
4. MATHON, C. C. and M. STROUN 1960. *Lumière et floraison (le photoperiodisms)*. Presses Universitaires de France, number 897. pp. 128.
5. MELCHERS, GEORG (editor) 1961. *Encyclopedia of Plant Physiology* XVI *External Factors Affecting Growth and Development*. Springer-Verlag, Berlin-Göttingen-Heidelberg. 950 pp.
6. MURNEEK, A. E. and R. O. WHYTE (editors) 1948. *Vernalization and Photoperiodism*. Chronica Botanica, Waltham, Mass. 196 pp.
7. VAN DER VEEN, R. and G. MEIJER 1959. *Light and Plant Growth*. The Macmillan Co., New York. 161 pp.
8. WITHROW, R. B. (editor) 1959. *Photoperiodism and Related Phenomena in Plants and Animals*. American Association for the Advancement of Science, Washington, D.C. 903 pp.

REVIEW ARTICLES ON FLOWERING, TIMING, OR LIGHT EFFECTS

9. BONNER, JAMES and JAMES LIVERMAN 1953. Hormonal control of flower initiation. pp. 283–303 in: W. E. LOOMIS (editor) *Growth and Differentiation in Plants*. Iowa State College Press, Ames, Iowa.
10. BORTHWICK, H. A. and S. B. HENDRICKS 1960. Photoperiodism in plants. *Science* 132: 1223–1228.
11. BORTHWICK, H. A., S. B. HENDRICKS, and M. W. PARKER 1956. Photoperiodism. pp. 479–517 in: Alexander Hollaender (editor) *Radiation Biology III. Visible and Near-visible Light*. McGraw-Hill, New York.
12. BÜNNING, E. 1956. Endogenous rhythms in plants. *Annual Review of Plant Physiology* 7: 71–90.
13. CHOUARD, P. 1960. Vernalization and its relations to dormancy. *Annual Review of Plant Physiology* 11: 191–238.
14. DOORENBOS, J. and S. J. WELLENSIEK 1959. Photoperiodic control of floral induction. *Annual Review of Plant Physiology* 10: 147–184.

15. HAMNER, KARL. Endogenous rhythms in controlled environments. Symposium Paper No. 13, Symposium on Environmental Control of Plant Growth at Canberra, Australia, August, 1962. In press.

16. HAMNER, K. C. 1958. The mechanism of photoperiodism in plants. *Photobiology, Proc. 19th Annual Biology Colloquium.* Oregon State College. pp. 7–16.

17. HARTSEMA, ANNIE M. 1961. Influence of temperatures on flower formation and flowering of bulbous and tuberous plants. *Encyclopedia of Plant Physiology* 16: 123–167.

18. HENDRICKS, STERLING B. 1956. Control of growth and reproduction by light and darkness. *American Scientist* 44: 229–247.

19. HENDRICKS, S. B. and H. A. BORTHWICK. Control of plant growth by light. Symposium Paper No. 14, Symposium on Environmental Control of Plant Growth at Canberra, Australia, August, 1962. In press.

20. LANG, A. 1952. Physiology of flowering. *Annual Review of Plant Physiology* 3: 265–306.

21. LIVERMAN, J. L. 1955. The physiology of flowering. *Annual Review of Plant Physiology* 6: 177–210.

22. LOCKHART, JAMES A. 1961. Mechanism of the photoperiodic process in higher plants. *Encyclopedia of Plant Physiology* 16: 390–438.

23. MELCHERS, G. and A. LANG 1948. Die Physiologie der Blütenbildung. *Biol. Zeitblatt.* 67: 105–174.

24. NAPP-ZINN, KLAUS 1961. Vernalization and verwandte Erscheinungen. *Encyclopedia of Plant Physiology* 16: 24–75.

25. NAYLOR, AUBREY W. 1961. The photoperiodic control of plant behavior. *Encyclopedia of Plant Physiology* 16: 331–389.

26. NAYLOR, A. W. 1953. Reactions of plants to photoperiod. *Growth and Differentiation in Plants.* pp. 149–178.

27. NAYLOR, A. W. 1952. The control of flowering. *Scientific American* 186: 49–56.

28. NAYLOR, A. W. 1952. Physiology of reproduction in plants. *Survey of Biological Progress* 11: 259–300.

29. NITSCH, J. P. The mediation of climatic effects through endogenous regulating substances. Symposium paper No. 11, Symposium on Environmental Control of Plant Growth at Canberra, Australia, August, 1962. In press.

30. PARKER, M. W. and H. A. BORTHWICK 1950. Influence of light on plant growth. *Annual Review of Plant Physiology* 1: 43–58.

31. PURVIS, O. N. 1961. The physiological analysis of vernalization. *Encyclopedia of Plant Physiology* 16: 76–122.

32. SALISBURY, F. B. 1961. Photoperiodism and the flowering process. *Annual Review of Plant Physiology* 12: 293–326.

33. SALISBURY, F. B. 1958. The flowering process. *Scientific American* 198: 108–117.

34. WAREING, P. F. 1956. Photoperiodism in woody plants. *Annual Review of Plant Physiology* 7: 191–214.

35. WASSINK, E. C. and J. A. J. STOLWIJK 1956. Effects of light quality on plant growth. *Annual Review of Plant Physiology* 7: 373–400.

36. WITHROW, R. B. 1959. A kinetic analysis of photoperiodism. pp. 439–471 in R. B. WITHROW (editor), *Photoperiodism and Related Phenomena in Plants and Animals.* American Association for the Advancement of Science, Washington D.C.

37. ZEEVAART, JAN A. D. Climatic control of reproductive development. Symposium paper No. 16, Symposium on Environmental Control of Plant Growth at Canberra, Australia, August, 1962. In press.
38. ZEEVAART, JAN A. D. 1962. Physiology of flowering. *Science* 137: 723–731.

LITERATURE CITED

39. BAUER, LEOPOLD and HANS MOHR 1959. Der Nachweis des reversiblen Hellrot-Dunkelrot-Reaktionssystems bei Laubmoosen. *Planta* 54: 68–73.
40. BENSON-EVANS, KATHRYN 1961. Environmental factors and bryophytes. *Nature* 191: 255–260.
41. BOGORAD, L. and W. J. MCILRATH 1960. Effect of light quality on axillary bud development in *Xanthium*. *Plant Physiology* 35: xxxii (Abstract in Supplement).
42. BONNER, JAMES and JAN A. D. ZEEVAART 1962. Ribonucleic acid synthesis in the bud, an essential component of floral induction in *Xanthium*. *Plant Physiology* 37: 43–49.
43. BORGSTRÖM, GEORG 1939. Formation of cleistogamic and chasmogamic flowers in wild violets as a photoperiodic response. *Nature* 144: 514–515.
44. BORTHWICK, H. A. and N. J. SCULLY 1954. Photoperiodic responses of hemp. *Botanical Gazette* 116: 14–29.
45. BURR, G. O., C. E. HARTT, H. W. BRODIE, T. TANIMOTO, H. P. KORTSCHAK, D. TAKAHASHI, F. M. ASHTON, and R. E. COLEMAN 1957. The Sugarcane Plant. *Annual Review of Plant Physiology* 8: 275–308.
46. CARR, D. J. 1957. On the nature of photoperiodic induction. IV. Preliminary experiments on the effect of light following the inductive long-dark period in *Xanthium pennsylvanicum*. *Physiologia Plantarum* 10: 249–265.
47. CUMMING, BRUCE G. 1959. Extreme sensitivity of germination and photoperiodic reaction in the genus *Chenopodium* (Tourn.) L. *Nature* 184: 1044–1045.
48. VON GAERTNER, T. and E. BRANNROTH 1935. Über den Einfluss des Mondlichtes auf den Blühterm in der Lang- und Kurztag- Pflanzen. *Bot. Centbl.*, Beihefte, Abt. A 53: 554–563.
49. HARDER, RICHARD 1948. Vegetative and reproductive development of *Kalanchoë blossfeldiana* as influenced by photoperiodism. *Symposium Society Experimental Biology* (Cambridge) No. II (*Growth*): 117–138.
50. HAUPT, WOLFGANG and EKKEHARD SCHÖNBOHM 1962. Das Wirkungsspektrum der negativen Phototaxis des *Mougeotia*-Chloroplasten. *Naturwiss.* 49: 42.
51. HIGHKIN, H. R. 1955. Flower promoting activity of pea seed diffusates. *Plant Physiology* 30: 390–391.
52. HUANG, RU-CHIH C. and JAMES BONNER 1962. Histone, a suppressor of chromosomal RNA synthesis. *Proceedings of the National Academy of Sciences* 48: 1216–1222.
53. JORAVSKY, DAVID 1962. The Lysenko affair. *Scientific American* 207 (Nov., No. 5): 41–49.
54. KOHLBECKER, RUTH 1957. Die Abhängigkeit des Längenwachstums und der phototropischen Krümmungen von der Lichtqualität bei Keimwurzeln von *Sinapis alba*. *Zeitschrift für Botanik* 45: 507–524.

55. LANG, ANTON 1958. Induction of reproductive growth in plants. pp. 126–139 in W. J. NICKERSON (editor), *Proceedings International Congress of Biochemistry, Fourth Congress, Vienna, 1958,* 6 (1958) *Biochemistry of Morphogenesis.* Pergamon Press, London.

56. LANG, ANTON and ERNST REINHARD 1961. Gibberellins and flower formation. *Advances in Chemistry* series. 28: 71–79.

57. LINCOLN, R. G., D. L. MAYFIELD, and A. CUNNINGHAM 1961. Preparation of a floral initiating extract from *Xanthium*. *Science* 133: 756.

58. LINCOLN, R. G., K. A. RAVEN, and K. C. HAMNER 1958. Certain factors influencing expression of the flowering stimulus in *Xanthium*. II. Relative contribution of buds and leaves to effectiveness of inductive treatment. *Botanical Gazette* 119: 179–185.

59. LOCKHART, JAMES A. and KARL C. HAMNER 1954. Partial reactions in the formation of the floral stimulus in *Xanthium*. *Plant Physiology* 29: 509–513.

60. LONGMAN, K. A. and P. F. WAREING 1959. Early induction of flowering in birch seedlings. *Nature* 184: 2037.

61. LÖRCHER, LARS 1958. Die Wirkung verschiedener Lichtqualitäten auf die endogene Tagesrhythmik von *Phaseolus*. *Zeitschrift für Botanik* 46: 209–241.

62. MAUNEY, J. R. and L. L. PHILLIPS 1963. The influence of day-length and night temperature on flowering of *Gossypium*. *Botanical Gazette* 124, March. In press.

63. MICHNIEWICZ, MARIAN and ANTON LANG 1962. Effect of 9 different gibberellins on stem elongation and flower formation in cold-requiring and photoperiodic plants grown under non-inductive conditions. *Planta* 58: 549–563.

64. MILLER, JOHN H. and PAULINE M. MILLER 1961. The effect of different light conditions and sucrose on the growth and development of the gametophyte of the fern, *Onoclea sensibilis*. *American Journal of Botany* 48: 154–159.

65. MOHR, HANS 1962. Primary effects of light on growth. *Annual Review of Plant Physiology* 13: 465–488.

66. MOONEY, H. A. 1961. Comparative physiological ecology of arctic and alpine populations of *Oxyria digyna*. *Ecological Monograms* 31: 1–29.

67. NITSCH, J. P. and F. W. WENT 1959. The induction of flowering in *Xanthium pennsylvanicum* under long days. pp. 311–314 in: R. B. Withrow (editor), *Photoperiodism and Related Phenomena in Plants and Animals.* American Association for the Advancement of Science, Washington, D.C.

68. SALISBURY, FRANK B. 1959. Growth regulators and flowering. II. The cobaltous ion. *Plant Physiology* 34: 598–604.

69. SCHWABE, W. W. 1959. Studies of long-day inhibition in short-day plants. *Journal of Experimental Botany* 10: 317–329.

70. SEARLE, NORMAN E. 1961. Persistence and transport of flowering stimulus in *Xanthium*. *Plant Physiology* 36: 656–662.

71. SHAPIRO, SEYMORE 1958. The role of light in the growth of root primordia in the stem of the Lombardy poplar. pp. 445–465 in: K. V. Thimann (editor), *The Physiology of Forest Trees.* The Ronald Press, New York.

72. STOKES, PEARL and K. VERKERK 1951. Flower formation in Brussels sprouts. *Mededelingen van de Landbouwhogeschool te Wageningen/Nederland* 50: 141–160.

73. STOUTEMYER, V. T., O. K. BRITT and J. R. GOODWIN 1961. The influence of chemical treatments, understocks, and environment on growth phase changes and propagation of *Hedera canariensis*. *American Society for Horticultural Science* 77: 552–557.

74. TAKIMOTO, ATSUSHI and KATSUHIKO IKEDA 1960. Studies on light controlling flower initiation of *Pharbitis Nil*. VI. Effect of natural twilight. *Botanical Magazine of Tokyo* 73: 175–181.

75. THOMPSON, P. A. and C. G. GUTTRIDGE 1960. The role of leaves as inhibitors of flower induction in strawberry. *Annals of Botany N.S.* 24: 482–490.

76. WAREING, P. F. 1959. Problems of juvenility and flowering in trees. *Journal of the Linnean Society of London Botany*. 366: 282–289.

77. WITHROW, ROBERT B. and LEONARD PRICE 1957. A darkroom safelight for research in plant physiology. *Plant Physiology* 32: 244–248.

78. DE ZEEUW, D. 1957. Flowering of *Xanthium* under long-day conditions. *Nature* 180: 588.

79. ZEEVAART, JAN A. D. 1962. DNA multiplication as a requirement for expression of floral stimulus in *Pharbitis Nil*. *Plant Physiology* 37: 296–304.

80. ZEEVAART, JAN A. D. 1962. The juvenile phase in *Bryophyllum daigremontianum*. *Planta* 58: 543–548.

81. ZEEVAART, JAN A. D. and ANTON LANG 1962. The relationship between gibberellin and floral stimulus in *Bryophyllum daigremontianum*. *Planta* 58: 531–542.

AUTHOR INDEX

ORGANISM INDEX

Scientific or common names are cited according to usage in the text, but all common plant names also show the Genus name.

SUBJECT INDEX

229